High Wired

On the Design, Use, and Theory of Educational MOOs

Cynthia Haynes and Jan Rune Holmevik, Editors

Foreword by Sherry Turkle

Ann Arbor

THE UNIVERSITY OF MICHIGAN PRESS

Published in the United States of America by
The University of Michigan Press
Manufactured in the United States of America
Printed on acid-free paper
2001 2000 1999 1998 4 3 2 1

A CIP catalog record for this book is available from the British Library.

Library of Congress Cataloging-in-Publication Data

High wired : on the design, use, and theory of educational MOOs /
Cynthia Haynes and Jan Rune Holmevik, editors ; foreword by Sherry
Turkle.

p. cm.

ISBN 0-472-09665-6 (alk. paper). — ISBN 0-472-06665-X (pbk.)

1. Education—Computer network resources. 2. Internet (Computer
network) in education. 3. Object-oriented databases. I. Haynes,
Cynthia. II. Holmevik, Jan Rune.

LB1044.87.H54 1998

025.06′37—dc21 97-52705

CIP

High Wired

Acknowledgments

This book is the result of a collaborative effort by the editors and other people from the United States and Norway. To all of our authors we are greatly indebted for supporting our vision and helping us make this book possible. Our deepest and foremost thanks, therefore, goes to John F. Barber, Jorge R. Barrios, Mark Blanchard, Amy Bruckman, Juli Burk, Brian Clements, Eric Crump, Pavel Curtis, D. Diane Davis, Jeffrey R. Galin, Dene Grigar, Michael Joyce, Beth Kolko, Ken Schweller, Sherry Turkle, Victor J. Vitanza, Shawn P. Wilbur, and Deanna Wilkes-Gibbs. It has been a great honor and privilege for us to work with all of you on this book.

We also want to sincerely thank our editor LeAnn Fields and the rest of the staff at University of Michigan Press. We could not have had a better and more supportive publisher.

Thanks are also due to our many friends from the MOO world whose generous help and support over the years have helped us understand and use this fascinating technology. In particular we send a heartfelt thank-you to Knekkebjoern (Jorge Barrios), Sky (Deanna Wilkes-Gibbs), Ringer (Isabel Danforth), Lord Chaos (Rui Mendes), Janus (Richard Godard), and Gustavo (Gustavo Glusman).

We must also thank the University of Texas at Dallas and the Department for Humanistic Informatics at the University of Bergen, Norway, for invaluable institutional and financial support.

Finally, we want to direct a very special thanks to Pavel Curtis and all the other people at LambdaMOO who made this wonderful technology that has had such a profound bearing on the course of our careers and lives, and to Amy Bruckman for building MediaMOO, the place where we met for the first time and started our collaboration on May 9, 1994.

Contents

Foreword: All MOOs are Educational—the Experience of "Walking through the Self"

Sherry Turkle

MUDs and MOOs

The term *virtual reality* is often used to denote metaphorical spaces within the computer that people navigate by using special hardware—specially designed helmets, bodysuits, goggles, and gloves. The hardware turns the body or parts of the body into pointing devices. For example, a hand inside a data glove can point to where you want to go within virtual space; a helmet can track motion so that the scene shifts depending on how you move your head. In a kind of virtual community known as a multi-user domain (MUD), instead of using computer hardware to immerse themselves in a vivid world of sensation, users immerse themselves in a world of words. Although some MUDs have some graphical component, MUDs are a text-based, social virtual reality.[1]

There are two basic types of MUDs. The adventure type, most reminiscent of the genre's heritage in the world of fantasy gaming, is usually built around a medieval landscape. In these MUDs, affectionately known by their participants as "hack and slay," the object is to gain "experience points" and increase one's powers. The jobs at hand are solving puzzles, slaying monsters and dragons, and finding gold coins, amulets, and other treasure. A second type consists of relatively open virtual spaces in which you can play at whatever captures your imagination. In these "social" MUDs, the point is to interact with other players and, on some MUDs, to help build the virtual world by creating one's own objects and architecture. "Building" on MUDs is something of a hybrid between computer programming and writing fiction. One describes a hot tub and deck in a MUD with words, but some formal coded description is required for the deck to exist in the MUD as an extension of the adjacent living room and for characters to be able to "turn the hot tub on" by pushing a specially marked "button." Building is made particularly easy in the type of social MUD that is the subject of this book. These are MOOs (MUDs of the object-oriented variety). In the past few years the number of social MOOs has exploded, as has the number specifically devoted to education, experimentation, and professional life. In this volume Jeffrey Galin includes over sixty in his annotated list, and every week brings announcements of new beginnings.

In the MOO world, characters communicate by invoking commands that cause text to appear on each other's screens. If I log onto LambdaMOO (one of the first and most popular MOOs, "written" and founded by contributor Pavel Curtis) as a male character named Turk and strike up a conversation with a character named Dimitri, the setting for our conversation will be a room in which a variety of other characters might be present. If I type 'Say hi', my screen will flash, You say "hi," and the screens of the other players in the room (including Dimitri) will flash, Turk says "hi." If I type 'Emote whistles happily', all the players' screens will flash, Turk whistles happily. Or I can address Dimitri alone by typing, 'Whisper Dimitri Glad to see you', and only Dimitri's screen will show, Turk whispers "Glad to see you." People's impressions of Turk will be formed by the description I will have written for him (this description will be available to all players on command), as well as by the nature of Turk's conversation and behavior.

In MOOs, virtual personae thus converse with each other, exchange gestures, express emotions, win and lose virtual money, and rise and fall in social status. For each MOO comes to have a unique social life and social structure. This is all achieved through writing, and this in a culture that has apparently fallen asleep in the audiovisual arms of television.

The contributing authors of this book are a wide-ranging group of scholars who have been drawn to MUDs and then more particularly to MOOs because of their educational possibilities. Most stress their vocation as a new venue for collaboration. Indeed, Jorge R. Barrios and Deanna Wilkes-Gibbs joke that their chapter is "just one small example of what two people nine time zones apart can do with only the MOO for real-time communication." Juli Burk was drawn to the MOOs' potential as theatrical environments. For Dene Grigar and John Barber, they were a place to reinvent the dissertation defense as a more public, interactive, and more inclusive rite of academic passage. But although their reports from the field are diverse in their emphases, there are commonalities. Multi-user virtual environments emerge as places where the boundary between the real and unreal is tested and redefined, where students and teachers can better learn to navigate between the physical and the virtual to create a new real that encompasses both.

Almost all of the authors stress the degree to which MOOs are a new form of writing. Cynthia Haynes and Jan Rune Holmevik point out that MOOS are *architextural,* that they create a unique and compelling genre: a landscape of words. This new writing is thus a kind of hybrid: it is landscape and it is momentarily frozen speech. In this new writing, unless it is printed out on paper, a screenful of flickers soon replaces the previous screen. The writing is dynamic, a living, of-the-moment trace. And yet it can be "logged" or "captured" as a transcript.

Michael Joyce, discussing the grounding of MOOs in writing, reminds us of what that implies when he cites Hélène Cixous: Writing, for Cixous, is "starting off," "walking through the self." In order to facilitate this discovery of self Joyce believes that the MOO, as an "artificial" environment, would do well to "disappear." The software should not be intrusive: what should remain is the social and educational space that it leaves as its "trace." The builder of a MUD is thus faced with a very particular design task: to build it right, build it so that it disappears. Jan Rune Holmevik and Mark Blanchard capture this paradox for the MUD builder or "wizard" when they quote from *The Wizard of Oz:*

> "I am Oz, the Great and Terrible. Why do you seek me?"
> Dorothy asked, "Where are you?"
> "I am everywhere," answered the Voice, "but to the eyes of common mortals I am invisible."

Eric Crump's essay on using text-based MUDs as writing centers makes the point that not only can the "walls" of a MUD disappear, but within a MUD, the very feeling that one is writing can disappear as well. Writing becomes conversation, fluent and charged with emotion. Yet the presence of a transcript of this spoken writing means that what has transpired can be edited, transformed into more conventional writing. This opens a path to writing for people who have always thought of themselves as incapable of it. On a MUD, one discovers that one can speak writing, something like Molière's M. Jourdain, who discovered that he had been speaking prose all his life. The writing on MUDs, because it is spontaneous and alive, is often rich in texture and emotion. Unconstrained and unconventional, it is a writing through which one encounters the self.

MOOs blur the boundaries between self and game, self and role, self and simulation. One player says, "You are what you pretend to be, you are what you play." But people don't just become whom they play, they play who they are or who they want to be or who they don't want to be. Players can use their MUD personae to refine their sense of who they are. Beth Kolko recasts the experience of a MOO in the language of travel. We discover who we are by visiting strange places. In MOOs, we visit with our virtual bodies, but they are no less places of "contact and conflict, both with the self and with others." They are places where we can learn and change because on them "the collisions of worldviews, of language, of cultural expectations are always in motion." As Amy Bruckman observes,

MUDs are "identity workshops." And in MUDs, identity building begins in an experience of play.

Identity Play

Play has always been an important aspect of our individual efforts to build identity. The psychoanalyst Erik Erikson (*Childhood* 52) called play a "toy situation" that allows us to "reveal and commit" ourselves "in its unreality." On a MUD one gets to build personae who live within a toy situation. In this way, a MUD can become a context for discovering who one is or wishes to be. This idea was well captured by one player who said:

> You can be whoever you want to be. You can completely redefine yourself if you want. You can be the opposite sex. You can be more talkative. You can be less talkative. Whatever. You can just be whoever you want, really, whoever you have the capacity to be. You don't have to worry about the slots other people put you in as much. It's easier to change the way people perceive you, because all they've got is what you show them. They don't look at your body and make assumptions. They don't hear your accent and make assumptions. All they see is your words. And it's always there. Twenty-four hours a day you can walk down to the street corner and there's gonna be a few people there who are interesting to talk to, if you've found the right MUD for you.

The anonymity of most MUDs (one is often known only by the name one gives to one's persona) provides ample room for individuals to express unexplored parts of themselves. A twenty-six-year-old clerical worker says, "I'm not one thing, I'm many things. Each part gets to be more fully expressed in MUDs than in the real world. So even though I play more than one self on MUDs, I feel more like 'myself' when I'm MUDding." In real life, this woman sees her world as too narrow to allow her to manifest certain aspects of the person she feels herself to be. Creating screen personae is thus an opportunity for self-expression, leading to her feeling more like her true self when decked out in an array of virtual masks.

For many of the people who use them, MUDs provide what Erikson called a psychosocial moratorium.[2] The notion of moratorium was a central aspect of Erikson's theories about adolescent identity development. Although the term implies a time out, what Erikson had in mind was not withdrawal. On the contrary, the adolescent moratorium is a time of intense interaction with people and ideas. It is a time of passionate friendships and experimentation. The moratorium is not on significant experiences but on their consequences. Of course, there are never human actions that are without consequence, so therefore there is no such thing as a pure moratorium. Reckless driving leads to teenage deaths, careless sex to teenage pregnancy. Nevertheless, during the adolescent years, people are generally given permission to try new things. There is a tacit understanding that they will experiment. Though the outcomes of this experimentation can have enormous consequences, the experiences themselves feel removed from the structured surroundings of one's normal life. The moratorium facilitates the development of a core self, a personal sense of what gives life meaning. This is what Erikson called identity.

Erikson developed his ideas about the importance of a moratorium to the development of identity during the late 1950s and early 1960s. At that time, the notion corresponded to a common understanding of what "the college years" were about. Today, thirty years later, the idea of the college years as a "time out" seems remote, even quaint. College is preprofessional, and AIDS has made sexual experimentation a potentially deadly game. But if our culture no longer offers an adolescent moratorium, virtual communities do. They offer permission to play, to try things out. This is part of what makes them attractive.

Erikson saw identity in the context of a larger theory of development in stages. Identity was one stage, intimacy and generativity were others. Erikson's ideas about stages did not suggest rigid sequences. His stages describe what people need to achieve before they can easily move ahead to another developmental task. For example, Erikson pointed out that successful intimacy in young adulthood is difficult if one does not come to it with a sense of who one is. This is the challenge of adolescent identity building. In real life, however, people frequently move on with incompletely resolved stages, simply doing the best they can. They use whatever materials they have at hand to get as much as they can of what they have missed. MUDs are striking examples of how technology can play a role in these dramas of self-repair. They are not a panacea; but they do present new opportunities.

Once we put aside the idea that Erikson's stages describe rigid sequences, we can look at the stages as modes of experience that people work on throughout their lives. Thus, the adolescent moratorium is not something people pass through but a mode of experience necessary throughout functional and creative adulthoods. We take vacations to escape not only from our work but from our habitual social lives. Vacations are times during which adults are allowed to play. Vacations give a finite structure to periodic adult moratoriums. Time in cyberspace reshapes the notion of vacation and moratoriums, because they may now exist in always-available windows. Erikson (*Childhood* 222) wrote that "the playing adult steps sideward into another reality; the playing child advances forward into new stages of mastery." In MUDs, we can do both; we enter another reality and have the opportunity to develop new dimensions of self-mastery.

On the computer bulletin board system known as the WELL, a long-lived virtual community based in San Francisco, an electronic discussion of online personae circled around the fact that experiences on the MUD permit a helpful externalization of one's conflicts. In other words, the presence of a record that you can scroll through again and again may alert you to your vulnerabilities or defensive reactions. In a certain sense, all computer applications can be projective screens. One thirteen-year-old I interviewed said: "When you use a computer and you put a little piece of your mind into the computer's mind and you come to see yourself differently." In the MUD environment, the creation of an avatar, a part of oneself, proceeds very much in this spirit. One New York City writer said ruefully, "I would see myself on the screen and say, 'There I go again.' I could see my neuroses in black and white. It was a wake-up call."

Identity and Multiplicity

When people adopt an online persona they cross a boundary into highly charged territory. Some feel an uncomfortable sense of fragmentation, some a sense of relief. Some sense the possibilities for self-discovery, even self-transformation. Serena, a twenty-six-year-old graduate student in history, says, "When I log on to a new MUD and I create a character and know I have to start typing my description, I always feel a sense of panic. Like I could find out something I don't want to know." Arlie, a twenty-year-old undergraduate, says, "I am always very self-conscious when I create a new character. Usually, I end up creating someone I wouldn't want my parents to know about. It takes me, like, three hours. But that someone is part of me."

One forty-two-year-old nurse whose real name is Annette calls herself Bette when she is online. "Annette," she says, "for all my life that will be sweet, little perky Annette from the *Mickey Mouse Club.* I want to be a Bette. Like Bette Davis. I want to seem mysterious and powerful. There is no such thing as a mysterious and powerful Mouseketeer." Online, Bette has an identity as a poet. "I've always wanted to write poetry. I have made little fits and starts through the years. I don't want to say that changing my name made it possible, but I can tell you it made it a whole lot easier. Bette writes poems. Annette just fools around with it. Annette is a nurse. Bette is the name of a writer, more moody, often more morose." When she types at her computer, Annette, who has become a skillful touch typist, says:

> I like to close my eyes and imagine myself speaking as Bette. An authoritative voice. When I type as Bette I imagine her voice. You might ask whether this Bette is real or not. Well, she is real enough to write poetry. I mean it's poetry that I take credit for. Bette gives courage. We sort of do it together.

Annette does not suffer from multiple-personality disorder. Bette does not function autonomously. Annette is not dissociated from Bette's behavior. Bette enables aspects of Annette that have not been easy for her to express. Becomeing more fluent as Bette, Annette moves flexibly between the two personae.

We could think of models of the self as falling along a continuum defined by two extremes: At one extreme, the unitary self maintains its oneness by repressing all that does not fit. At the other extreme is the person who suffers from multiple-personality disorder. This person's multiplicity exists in the context of an equally repressive rigidity because the parts of the self are not in easy communication. Communication is highly stylized; one personality must speak to another personality. If we think of the "disorder" in multiple-personality disorder as the need for the rigid walls between the selves (blocking the secrets those selves protect), then we can begin to construct a model of a healthy self that has fluid access among its many aspects. It is not a unitary self but a flexible self.[3] It is not a self of split-off parts (as in multiple-personality disorder) because it is easy to cycle through its multiple aspects, aspects that are themselves in constant communication with each other. Its essence is illuminated by the philosopher Daniel Dennett's "multiple drafts" theory of consciousness. Dennett analogizes the mind to the commonplace experience of having several versions of a document open on a computer screen. The writer, the computer user, is able to cycle through them at will. The presence of the multiple versions, each of which has certain virtues and certain limitations, encourages a respect for the different versions while it imposes a certain distance from them. The author does not completely identify with any one draft. Similarly, the open lines of communication among the aspects of the flexible self encourages an attitude of respect for the many within us and the many within others. No one aspect can be claimed as the absolute, true self.

The experience of creating MUD avatars increases our appreciation of our inner diversity as well as a healthy respect for the limitations of our self-knowledge. The historian of science Donna Haraway writes of this "split and contradictory self" in a positive way, going so far as to equate it with a "knowing self" (22). She is optimistic about its possibilities: "The knowing self is partial in all its guises, never finished, whole, simply there and original; it is always constructed and stitched together imperfectly; and *therefore* able to join with another, to see together without claiming to be another."

When identity was defined as unitary and solid, it was relatively easy to recognize and censure deviation from a norm. A more fluid sense of self allows a greater capacity for acknowledging diversity. It makes it easier to accept the array of our (and others') inconsistent personae—perhaps with humor, perhaps with irony. We do not feel compelled to rank or judge the elements of our multiplicity. We do not feel compelled to exclude what does not fit. In all of these ways, life in the MUD is educational in ways that go far beyond any specific MUD content—the very experience of building and playing a MUD persona is a powerful education.

Virtuality as Transitional Space

In an article that attempts to capture the possible therapeutic side to virtual reality, Leslie Harris speculates on how virtual experiences become part of the perceptual and emotional background that changes the way we see things. Harris describes an episode of *Star Trek: The Next Generation* in which Captain Picard plays Caiman, an inhabitant of the virtual world Catanh. On Catanh, Picard lives the experiences he had to forgo in order to make a career in Starfleet. He has a virtual experience of love, marriage, and fatherhood. He develops relationships with his community that are not possible for him as a Starfleet commander. Harris says, "He can eventually fall in love with a fellow crew member in his 'real life' because he experienced the feelings of love, commitment, and intimacy 'on' Catanh."

Here, virtuality is powerful, transitional, and put to the service of Picard's embodied real. Buddhists speak of their practice as a raft to get to liberation, waiting on another shore. But the raft, like the period of treatment in the psychoanalytic tradition, is thought of as a tool that must be set aside. Wittgenstein takes up a similar idea in the *Tractatus,* when he compares his work to a ladder that is to be discarded after the reader has used it to reach a new level of understanding. As an identity-building experience, MUDs are similar in their transitional dynamics.

In April 1995, a town meeting was held at MIT on the subject "Doing Gender on the Net." As the discussion turned to using virtual personae to try out new experiences, a thirty-year-old graduate student, Ava, told her story. She had used a MUD to try out being comfortable with a disability. Several years earlier, Ava had been in an automobile accident that left her without a right leg. During her recuperation, she began to MUD. "Without giving it a lot of advance thought," Ava found herself creating a one-legged character on LambdaMOO. Her character had a removable prosthetic limb. The character's disability featured plainly in her description, and the friends she made on the MUD found a way to deal with her handicap. When Ava's character became romantically involved, she and her virtual lover acknowledged the "physical" as well as the emotional aspects of the virtual amputation and prosthesis. They became comfortable with making virtual love, and Ava found a way to love her own virtual body. Ava told the group at the town meeting that this experience enabled her to take a further step toward accepting her real body. "After the accident, I made love in the MUD before I made love again in real life," she said. "I think that the first made the second possible. I began to think of myself as whole again." For her, the MUD had been a place of healing. On the MUD, she was able to reimagine herself not as whole but as whole-in-her-incompleteness. Each of us in our own way is incomplete. Virtual spaces may provide the safety for us to expose what we are missing so that we can begin to accept ourselves as we are.

Virtuality need not be a prison. It can be the raft, the ladder, the transitional space, the moratorium that eventually is discarded in order to reach greater freedom. We don't have to reject life on the screen, but we don't have to treat it as an alternative life either. We can use it as a space for growth. Having literally written our online personae into existence, we are in a position to be more aware of what we project into everyday life. Like the anthropologist returning home from a foreign culture, the voyager in virtuality can return to a real world better equipped to understand its artifices.

The papers in this volume, papers by educators, computer scientists, and systems developers who have immersed themselves in the world of MOOs, can be read as the collective musings of a group of people who have tried to get the best of both worlds. In many different ways they approach the question of how experience in the virtual helps us reflect on the embodied real. I have often referred to computers as evocative objects for our time, objects that cause us to reflect upon ourselves. In the course of these essays, MOOs are revealed as evocative objects for thinking about society, education, creativity, reality, and, perhaps most fundamentally, for thinking about the self.

The authors come upon this question in different ways. Victor Vitanza, writing in a philosophical language, raises the question of how the objects of our lives shape us as we shape them: "While I and my kind were folding paper into airplanes and spitballs, other kids in the Orient were being introduced to origami, the art of paper folding." Writing as a developer of virtual tools, Ken Schweller describes the new set of materials through which we will come to discover, or "unfold," ourselves as we discover our objects. Amy Bruckman, who developed MediaMOO, a virtual community for media researchers, highlights how we search the Net to "find ourselves" in cyberspace; Pavel Curtis writes about his own "coming of age" in a series of experiences that are indissociable from the maturation of a virtual community, a point Shawn Wilbur supports when he stresses that "for its administrators, every MOO is an educational MOO." Diane Davis alerts us to the sense in which some people tend to use the fact of virtual reality to reassure themselves that there is a "real" reality, but that the experience of a MOO calls us to question any such simple formulations. MOOs, it would seem, educate us about a future sense of the "real" by calling us to a new form of joint citizenship in its several aspects, an issue Brian Clements joins in his poetry when he asks: "Does the tightrope dancer stop dead when he thinks / of the dual nature of the rope?" And Michael Joyce, who cited Cixous's notion of writing as "walking through the self," reminds us of what draws together all of these works: the power of MOOing is essentially the power of facing ourselves and making ourselves in the mirror of language.

Notes

1. This foreword draws on my book *Life on the Screen,* where the notion of MUDs as a location for identity play is elaborated in fuller detail.
2. "The adolescent mind is essentially a mind of the *moratorium*" (Erikson, *Childhood* 262). Erikson's psychobiographies (*Young Man Luther, Gandhi's Truth*) illustrate the working out of the notion of the moratorium in individual lives.
3. The notion of a flexible self is evoked in many places. I have been particularly influenced by the work of Emily Martin, Kenneth Gergen, and Robert Jay Lifton.

Works Cited

Bruckman, Amy. "Identity Workshop: Emergent Social and Psychological Phenomena in Text-Based Virtual Reality." March 1992. Internet. Available: ftp://media.mit.edu/pub/asb/papers/identity-workshop.
Dennett, Daniel C. *Consciousness Explained.* Boston: Little, Brown, 1991.
Erikson, Erik H. *Childhood and Society.* 2d ed. New York: Norton, 1963.
———. *Gandhi's Truth: On the Origins of Militant Nonviolence.* New York: Norton, 1969.
———. *Young Man Luther: A Study in Psychoanalysis and History.* New York: Norton, 1958.
Gergen, Kenneth. *The Saturated Self: Dilemmas of Identity in Contemporary Life.* New York: Basic Books, 1991.
Haraway, Donna. "The Actors Are Cyborg, Nature Is Coyote, and the Geography Is Elsewhere: Postscript to 'Cyborgs at Large.'" In *Technoculture,* ed. Cons-

tance Penley and Andrew Ross. Minneapolis: University of Minnesota Press, 1991.

Harris, Leslie. "The Psychodynamic Effects of Virtual Reality." *Arachnet Electronic Journal on Virtual Culture* 2.1. February 1994. Internet. Available: ftp://mbdu04.redc.marshall.edu/pub/ejvc/HARRIS.V2N1.

Lifton, Robert Jay. *The Protean Self: Human Resilience in a Postmodern World.* New York: Basic Books, 1993.

Martin, Emily. *Flexible Bodies.* Boston: Beacon Press, 1994.

Turkle, Sherry. *Life on the Screen: Identity in the Age of the Internet.* New York: Simon and Schuster, 1995.

Introduction: "From the Faraway Nearby"

Cynthia Haynes and Jan Rune Holmevik

> I suppose I am odd but I do like the far away—With all the earth being rearranged as it is these days I sometimes wonder if I am crazy to walk off and leave it and sit down in the far away country as I have—at least it is quiet here—I have no radio—and no newspaper—and a feeling of much space.
>
> —Georgia O'Keeffe

It is a rare and radiant chance to write from New Mexico and Norway in the span of a few months, but our collaboration has been rare and radiant. So it seems fitting to have conceived of this collection of essays while in the "faraway nearby" of O'Keeffe country, sitting in the modest library of the Ghost Ranch conference center, miles away from the nearest Internet provider. Now we sit facing the Geiranger Fjord in a remote and breathtaking area of western Norway. From the faraway nearby we gather the digital drafts of our colleagues and read them beside glacial lakes and snow-capped mountains shrouded in mist that dissolves with occasional bursts of sun. The paradox of being out of range of Internet culture is not lost on us. In fact, *High Wired* is a compilation that defies orthodoxy, that takes paradox as its genre.

Our aim, from the beginning, has been to put together a collection of disparate and diacritical essays on educational MOOs (multi-user dimension, or MUD, object-oriented). We sought to pry open the genres of technical handbooks and theoretical reference texts and to splice them into something different, brought together in the "radiant space" of the "faraway nearby" with all the courage of poetry. Indeed, each section of *High Wired* opens with a poem, and we think the result is like drinking New Mexican sunsets under the Norwegian midnight sun—it is like nothing else.

Educational MOOs may or may not fall in the category of distance learning. This depends on whom you ask, but they certainly represent a trend to technologize education that is escalating. Not only do MOOs connect students locally, they connect students at a global level, with each other and with resources otherwise inaccessible. What schools and universities may not know is that they can host an educational MOO and operate it for users of all platforms and systems for a fraction of the cost of other kinds of Internet-based technology. *High Wired* aims to negotiate the era of tight budgets and tight schedules, of the infoglut of tight rhetorics about the Internet. In other words, our book takes a practical and theoretical turn (or rhetorical trope) toward the needs of educators.

MOOs in Context

The history of educational MOOs is essentially a story about adaption and reconception of gaming technology for professional and educational use. The story is still very much in progress. The complete story of MUDs is beyond the scope of this book, but some background on the origins of this technology and how it has evolved may be useful.

MOOs evolved from MUDs. The first MUD was written in Great Britain in 1979 by Roy Trubshaw and Richard Bartle, students at Essex University. It was a small multiplayer game in which you could move between locations and chat with other players in real time. The object of the game was to explore a virtual fantasy world, find treasures, and kill monsters, until you achieved the status of wizard. Due to its accessibility via computer networks, the game became quite popular, and copies of it quickly spread to Norway, Sweden, Australia, the United States, and other countries.

The next installment in the prehistory of MOOs began ten years later with the development of a new type of MUD known as TinyMUD. The original TinyMUD was written in 1989 by James Aspnes, then a graduate student at Carnegie-Mellon University. In contrast to the traditional MUDs with their Dungeons and Dragons–inspired themes, the TinyMUD was conceived of as a player-extendable place where the main focus was on social interaction (players designed the space to resemble rooms, houses, castles, hotels, etc.). Although the original TinyMUD lasted less than a year, several other TinyMUD-type systems were developed in the late 1980s and early 1990s. Perhaps the most important new feature of these systems was their player-programmability. Players could now not only add to the world by building new rooms and other objects, but they could write programs that altered the MUD's universe in much more profound ways than what was possible before. One of these programmable MUDs developed on the basis of TinyMUD was MOO. It was originally written by Stephen White, but the development was later taken over by Pavel Curtis, at the time a computer science researcher at the Xerox Palo Alto Research Center (PARC). Curtis made several enhancements to the server, and in the fall of 1990 he started his own MOO, LambdaMOO. As Curtis explains in his essay, LambdaMOO quickly became a popular hangout on the Internet. It has been running continuously for over seven years and is likely the biggest MUD on the Internet, with thousands of players from all over the world.

Some of the early visitors to LambdaMOO realized that this technology had a great potential for more professional applications. Amy Bruckman, then a graduate student of MIT's media lab, envisioned the MOO as a virtual meeting place for media researchers, and in 1992 she started MediaMOO, which has had an enormous significance for the professionalization of MOO technology. It demonstrated that a technology that had predominantly been used for gaming and online chat purposes could be put to serious and productive use. Second, it became an important meeting place for academics and others curious about this technology and what it could be used for. Like the editors and some of the contributors of this book, lots of people have met at MediaMOO and formed working relationships that would not otherwise have been possible.

After the establishment of MediaMOO, other professionally oriented MOOs began to appear. BioMOO, founded by Gustavo Glusman and Jaime Prilusky of the Weizmann Institute of Science in Israel in 1993, was built as a meeting place for biologists. About the same time, Jeanne MacWhorter, a social worker, had the idea of starting a MOO for people of her profession. As she began work on this project, however, she realized that it would be more fruitful to incorporate other disciplines, and thus was born Diversity University, the largest multidisciplinary MOO on the Internet today.

With MediaMOO, BioMOO, AstroVR, and other professional MOOs of the first generation, educators could not bring their classes online. Recently, however, MOOs with more explicit educational missions have appeared, where educators are welcome to bring their students online. Here they explore the potential of this technology for a host of applications such as online writing centers, electronic classrooms, hypertextual writing spaces, Net-based collaborative environments, or complete cyberspace-based campuses, as Diversity University is today.

Why This Book?

As cocreators and administrators of Lingua MOO, we have experienced firsthand the complicated task of setting up an educational MOO and administering its daily operation. We realized before long that others also found this task daunting given the technical knowledge required and the lack of centralized, accessible resources. In the past few years we have had inquiries from many people who have encountered obstacles in setting up their own educational MOO. Computer-mediated communication (CMC) literature often makes mention of MUDs and MOOs. It may go so far as to define what a MUD is and how MOOs evolved. But the literature stops short of providing the kinds of technical details that are needed, and the definitions lack applicatory contexts and theoretical explorations of the underlying implications for using MOOs in education. *High Wired* is designed to answer the most technical questions, while also entertaining the most abstract ideas. It is geared for practical use, handy reference, and guided study. Teachers, students, administrators, and all users new to MOO will find something beneficial inside.

We have designed several complements to the book: a World Wide Web site for the project, complete with updated information on technology changes; a *free* MOO core called the *High Wired* enCore developed by our authors with many of the educational tools we refer to in the book; and finally, a special site in Lingua MOO that represents the book as "text" and "cyphertext."[1]

The Architexture of MOOs

As an educational space, the MOO is not just a place to manipulate code, text, and the designs of engineers or programmers but can be, as our authors show, a space for *bricoleurs,* for those who create from what they have on hand, constructing and recycling imaginative murmurings in archi*textual* ways. MOOs reinvent the notion of education, and their users reconceive this space to accommodate radically different genres of discourse and pedagogies. We see our efforts to move beyond the gaming history and rigid understandings of MOO as a lever with which to pry open current notions of MOOspace and the discourse that constitutes it. One way to do that is to give several conceptual starting points toward new definitions of MOO.

On the surface, perhaps the most distinct feature of the MOO is that it is all text. Compared to the fancy graphics, real audio, live video, cool talk, and silly ideographic avatars that you can find on the World Wide Web these days, such an interface may seem antiquated. However, work is under way now to give MOOs a truly multimedia interface. The three-dimensional nature of most MOOs is still text-based, by design: in fact, it is the simple text itself that makes MOO an intriguing and powerful space.

Conventional MOO architecture consists of two basic components: a server that handles input and output as well as the multiple connections with logged-in users, and a database, which is the MOO itself, containing the digital representation of the virtual world. Within this system the object-oriented structure easily allows MOO programmers and builders to extend the database in creative ways. Since the system was derived from adventure games like Dungeons and Dragons and Zork, the architecture has replicated real-life dungeons, caves, and rooms.

Ethics and Educational MOOs

If the architecture of MOOs replicates real-life places, we can be assured that real-life problems exist there, too. Educators need to be aware of how and in what forms these problems take shape online, in addition to knowing how to handle them. It is not easy to find answers in the mountains of popular-culture hype and mass-media reports on the Internet. Depending on what magazines and newspapers you read, Internet technology is either the salvation or the demise of civilization. Certainly the rapid growth of access to the Internet has parents and educators wringing their hands over how to protect children and students online from some of the same dangers they face in the physical world. Yet one of the most encouraging aspects of the growth and scope of the Internet is the opportunity for all of us to contribute to the construction of ethical guidelines in the development and use of Internet-based technologies in educational settings.

Along with the opportunities, however, come responsibilities to practice "safe rhetoric," and that includes *not* discounting stories of specific abuses on the Internet, *not* embracing MOOs uncritically, at the same time *not* contributing to uninformed media hype surrounding actual cases of actual violence perpetrated with the aid of the Internet. For those of us who want to incorporate the Internet in our teaching, the conundrum comes when research lags behind pedagogy, often resulting in a kind of blind faith—the kind we all remember in the physical world when (as naive youth) we thought "I am invulnerable, nothing will happen to me."

Teachers can expect to encounter problems like plagiarism (either outright cheating or uneven collaboration), flaming (extremely rude or mean responses), Internet stalking, and other forms of harassment. Make no mistake, these are serious problems that warrant serious countermeasures. Just as you teach your students to be critical thinkers, that is, not to accept their world as given, you need to teach them not to accept virtual worlds as given either. We remember several incidents on our MOO (and there are countless others on other educational MOOs) that required swift action and discipline.

A teacher logged her class on to Lingua MOO but left the physical classroom once her students were happily (or so she thought) discussing their reading assignment. Within the Moo, we went to join the class to see if they needed help. Almost immediately, several students began to make inappropriate remarks to us. They also had chosen offensive guest names to use while online. We had to ask them to log off, and when that did not work, we ended up having to disconnect the whole class from the MOO. In that instance, the follow-up with the teacher was a lesson learned: never leave your class unless they are thoroughly familiar with the etiquette protocols and unless you are comfortable with the technology yourself so you know how to prevent this kind of problem or rectify it while it is happening. Teaching in real-time learning environments requires immediate action.

Teachers should also anticipate disruptions and take preventive action. For example, there may be times when other MOO players interrupt a class with what we call "drive-by shoutings," where they teleport into the virtual classroom and say silly or offensive things, or just stumble into the class innocently in the hope of chatting. But if teachers understand the various options for restricting access to the MOOspaces they use for holding class, they will be able to prevent these kinds of disruption. There are, however, problems that they cannot always prevent. Because it is possible to page private messages to users while logged into a MOO, teachers may never be aware of problems until it is too late, the damage done. In one instance, we recall how the MOO community worked together to resolve a serious harassment incident.

We were working on Lingua MOO one day, and a teacher friend of ours paged us from another MOO (we are networked to twenty other educational MOOs). She wanted to know how to handle a problem that a student was having online. The student had been the recipient of some offensive private pages and wanted to file a complaint with the MOO administrator. The MOO administrator was not online at the time, and the teacher did not know exactly how to handle the complaint. She brought the student to our MOO, we listened to what happened, and then we made suggestions about how to resolve the incident. The student felt better and took our advice. If this incident had happened in a traditional classroom setting, in all likelihood swift action and nearby expert advice would not have been possible. Teachers on the Internet have at their beck and call an amazing community of educators and professionals.

There is a fine line between giving students the latitude to be creative online and setting them loose with no ethical boundaries whatsoever. Just as in traditional classroom settings, common sense and proper preparation go a long way toward anticipating problems and resolving them. It is not necessary to blame the whole of cyberspace and thereby throw the virtual baby out with the virtual bathwater. There are ways to manage the fluidity associated with identity, speed, and open access on the Internet. We need to avoid the rhetoric of negative cyber-hype and engage instead in productive discussion about the Internet in terms of its positive impact on individual and collective lives. Especially with respect to the use of Internet technology in education, we should proceed rhetorically, going slowly, doing our homework, so to speak, before we make claims about the value of the Internet.

Educational MOOs have opened channels for efforts to resolve the kinds of dramatic circumstances mentioned above. The paradox of the Internet is that it is often perceived as both poison and remedy. But where ethics is concerned, we have all manner of rhetorical strategies at our disposal in virtual educational spaces. The question to consider now is whether we resort to scapegoating the Internet as the newest cause of our social ills, or whether we practice safe rhetoric in representing ourselves to the datasphere. We need to *understand,* to *stand under,* our own claims with rhetorical sensibility as we stand in our real and virtual classrooms. The very nodes that network students with a complex world, these links to passion and paradox, also allow us to see that even though we all share the same susceptibility to technologically transmitted dis-ease, there are more reasons than not to construct educational MOOs. Let us build these communities as text-based synchronous safe havens where learning is not sacrificed to appease the skeptical and where teaching in MOOspace offers amnesty to even the most technologically faint of heart.

How *High Wired* Unfolds

Part I of this book introduces the concepts and contexts of MOOs. Part II provides the practical and technical information needed to create and operate a MOO, and part III provides examples of how MOO technology is currently being used in academic settings, including online research and collaboration, teaching, conferencing, networking, and other useful applications. In part IV, theorists examine the complicated layers of identity, social interaction, community, and life that constitute MOOspace. Finally, the book's appendix provides a compendium of MOOs.

Amy Bruckman's opening essay considers the concept of virtual communities, how they relate to real-world communities, and how they can be shaped and constructed. The Net is made up of hundreds of thousands of separate communities, each with its own special character, but it is difficult to find a good virtual community—so much so that you often must start your own. This is precisely what Bruckman did in establishing MediaMOO in 1992. Bruckman is also the architect of another virtual community, MOOSE Crossing, designed to be a constructionist learning environment for children, and her essay suggests design techniques that help these neighborhoods develop a desired character. The challenge for the future, she contends, is not to purge the online world of unwholesome influences, but to develop neighborhoods that suit the needs of particular groups of people.

Pavel Curtis's essay provides a fascinating and highly personal recollection of his involvement with MOOs. Curtis is the founder and archwizard of LambdaMOO, the biggest and one of the most successful MUDs on the Internet, and is a major figure in the world of MOOs.

Part II opens with Deanna Wilkes-Gibbs and Jorge Barrios's essay, offering a detailed and informative introduction to the tools needed to use a MOO productively. Wilkes-Gibbs and Barrios are cofounders of Meridian, a MOO devoted to virtual travel, cultural exchange, and interaction. The chapter opens with a discussion of the prerequisites for using a MOO and how to log in for the first time. They then move on to an in-depth discussion of basic communication and navigation techniques in MOO, and then discuss the online editing facilities in the MOO. In the latter part of the chapter they touch briefly upon issues related to programming in MOOs, providing an introduction and starting point to further exploration of this highly interesting capability.

The next essay, by Ken Schweller, discusses educational tools that enhance teaching and writing productivity in the MOO, including tools for recording activities online, for making presentations, and for holding classes online. Schweller also discusses the design issues and considerations for making educational tools for the MOO environment.

Jan Rune Holmevik and Mark Blanchard's essay describes starting up a new MOO. After discussing the prerequisites for running a MOO, they explain how to download the necessary software, configure it, set up user accounts, and build the first few core locations. They discuss issues related to MOO administration and what needs to be considered in terms of security and daily operation.

The last essay in part II deals with day-to-day MOO administration. Shawn Wilbur, a longtime MOO administrator, considers issues involved in running a MOO. The essay is designed to help new administrators learn how to think about MOOs and the various tasks involved in maintaining a site. Issues such as planning, creation, and maintenance, in both their technical and social aspects, are covered in this essay.

Parts III and IV gather essays that offer a range of diverse applications of educational MOOs and more theoretical meditations on virtual reality, embodiment, MOO architecture, and momentariness. Opening part III, Cynthia Haynes's essay serves as a general guide for teaching with MOOs, stressing the importance of preparation and practice, and encouraging teachers to balance clearly articulated goals with creative assignments and activities. Haynes's experience as both MOO administrator and teacher helps illuminate problematic issues that range from technical questions about MOO clients to tips for handling inappropriate conduct online.

The remaining essays in part III offer unique applications of MOO that go beyond strictly classroom-oriented MOO activity. In Eric Crump's essay, "At Home in the MUD: Writing Centers Learn to Wallow," readers will find a witty foray into the nitty-gritty world of online tutoring and teaching. Crump depicts the benefits of online writing centers, yet he does not lose sight of the problems of such virtual interactions.

Another fascinating application of educational MOOs is examined in Dene Grigar and John F. Barber's essay, "Defending Your Life in MOOspace: A Report from the Electronic Edge." Based on Grigar's history-making online dissertation defense, held at Lingua MOO in 1995, the essay analyzes the event and its implications for future scholarship, in an innovative format, a collage of MOO dialogue, MOO slides, ASCII maps of the MOOspace, emails, and multivocal sections. As Grigar and Barber conclude, there can be problems with the use of this medium at the highest levels of academic work, which have to date resisted the integration of technology. They invite members of the academy to step out onto an "electronic edge" where technology and the university offer unlimited resources for making and marking new knowledge.

In theater scholar Juli Burk's essay, "The Play's the Thing: Theatricality and the MOO Environment," she argues that educational MOOs have moved away from their gaming heritage and can enable women to enact their subjective agency in radically new ways that resist patriarchal structures. Burk establishes the performative nature of MOOs by comparing them to traditional theatrical environments and offers her experience as creator and administrator of ATHEMOO, a MOO for theater scholars and educators. Burk's work indicates that educational MOOs are opening doors for people previously denied entrance into traditionally male-dominated domains of knowledge and power.

Part IV offers a series of theoretical studies that reinforce the notion that MOOs defy traditional understandings about reality, the "self," community, life and death. The subjects of these challenging essays are complex, and not easily summarized, but brief accounts of chapters written to resist such reduction may be useful. They invite readers to savor them, to revisit them often in order to trek with the *thought* of difficult themes.

In "(Non)Fiction('s) Addictions: A NarcoAnalysis of Virtual Worlds," Diane Davis writes a stunning and playful commentary on the "reality principle" that undergirds the "foundations of the academy." Davis enacts her excursus by taking a textual "swipe" at the conventional format of the essay. With irreverent humor, Davis scuttles the idea that the "real" is a mirror of reality. She cracks and shatters that mirror as she invokes the Energizer bunny to personify an "era of simulation" that keeps "going and going and going." Logging on to the MOO, she muses, is not putting on some virtual reality that disappears with the on/off switch—there *is* no on/off switch. As virtual life slides into real life, Davis challenges the distinctions that keep the question of education safely within the domain of the real.

Beth Kolko's essay, "Bodies in Place: Real Politics, Real Pedagogy, and Virtual Space," blurs still more boundaries. Taking the position that in the MOO "language and identity" work to create a fuzzy border between body and place, Kolko lifts her "subject" beyond conventional orders of representation and enters the "gap between the known and the unknown." In the MOO, Kolko argues, players may "bend social codes" and present themselves in ways that articulate "political choices," but the real issue for educators is how we can make this realization into "pedagogically productive work."

Victor Vitanza's essay, "Of MOOs, Folds, and Non-reactionary Virtual Communities," rocks the volume with its incendiary language. Vitanza lights the flame of justice amid a rousing romp through histories of reactionary violence he would have us unlearn in the learning spaces, folds, and bearings of the educational MOO. The question of rape in cyberspace operates as a founding narrative in Vitanza's "history/hystery" of virtual societies. He calls for a "refolding" of the concept of justice, a thinking our way out of reactionary principles that ground (by way of the "principle of sufficient reason") current notions of justice in the MOO and in the world. Readers are invited to savor, to study with care, Vitanza's thinking, and to move beyond the oppressive code of academic discourse and surface orientations, to lose our nostalgia for (the) surface.

"Words are the way, I want to say." The simplicity and grace with which hypertext novelist and theorist Michael Joyce utters a text into being is like gliding in a sailboat, our wind caught up in the sails of meaning, our eyes trained toward that point on the horizon where the water lifts up into the sky. Joyce spreads his landscape before us and shines particular light on the ideas of self, persistence, and the momentary. The MOO sparkles, for him, with the freshly dabbled paint of existences, of life amplified, amplifications lived.

Cypher/TEXT and *Élekcriture*

It is our goal to conceive new metaphors with which to play the *bricoleur,* to design the space with text as the primary metaphor *and* building blocks. This is what we call *cypher/TEXT,* a word that assembles in one term the notions of cyber, hyper, text, and most important of all, the reader herself. We think of it as a three-dimensional text, though not in the conventional physical definition of that phrase. While traditional text can be thought of as one-dimensional and linear, and hypertext as multidirectional and two-dimensional because of its ability to link documents, *cypher/TEXT* adds a third dimension by bringing the reader/writer actively *into* the text. In the MOO, the reader is represented through textual descriptions. You interact with others through textual dimensions, through textual ethos, pathos, and logos. Thus, readers speak, emote, and think in several dimensions, but more than that—they *are* a textual dimension in and of themselves.

We know that discussion on MOOs happens by means of a kind of oral writing, or orality put into writing (orality and writing being forms of discourse that have traditionally been juxtaposed and that have been the basis for centuries of conflict between philosophers and rhetoricians). The effects of this generate a different genre, one that breaks the barriers of space and time (not to mention distance), and that produces a unique and seamless fabric of architextures and assemblages, a different genre of writing. What makes this genre of discourse archi*textural* (and generative) is that discussants engage in real time, by writing text in a space that is *itself* textually assembled, or constructed, and performed by personae that are themselves textually constructed in descriptive and narrative forms, and who assume identities that may be equally constructed. The space is not only text based, but it breaks the boundaries of physical space by bringing together, or assembling, people across continents and time zones. Thus, perhaps for the first time, MOOs enable a fluidity among these factors; rather than a factory that produces no product, it is a generative space (or womb) capable of giving birth to endless streams of text, personae, and virtual spaces. Inspired by a French feminist perspective, we call this kind of writing *élekcriture,* a word that captures, or should we say *embodies,* the various factors we have just described. The word *elektra* comes from the Greek for "the beaming sun," and *écriture* is conceived as a French feminist writing that resists traditional hegemonic discourse as it seeks to assemble or construct new discourse. With *élekcriture* it is possible, provisionally, to represent all that is assembled and that which is constitutive of the archi*texture* of MOOs.

The challenge now is to augment the MOO database with new methods of assembly, that is, to begin the task of implementing *cypher/TEXT* MOOs and considering the implications of this new architecture (and these new metaphors) on the functionality of electronic learning spaces. For example, because of the often disconcerting nature of the *élekcriture,* we have created spaces that allow us to control the flow of discourse, like the auditorium at Lingua MOO where Dene Grigar defended her dissertation online in 1995 (see chapter 9). These types of spaces are successful insofar as they resemble real-life auditoriums, seminar rooms, classrooms, and so forth. Thus, new metaphors like *cypher/TEXT* and *élekcriture* meet the challenge of these variations.

Another metaphor that Victor Vitanza explores in his essay is based upon the notion of the "fold," or more specifically, the folded page. Vitanza is not, however, taking the traditional printcentric fold into his reconception; he is using Leibniz, Deleuze, the art of origami, and the unique nonstructure of the Möbius strip to reconceive the fold as a kind of "hypergami" (a term used by some mathematicians who have created educational software based on origami). In other words, when one finds oneself inside a fold in the MOO, one finds pages folded and refolded in omni-directional fashion, that is, folded in all directions at once. When we take into account the possibility of multiple users logged into such a space, these metaphors take on even more nuances than their concepts can hold, or the fold can conceive. Thus, the assembly of a *cypher/TEXT* space is different—it contains and is contained in all directions simultaneously and by as many variations as there are users logged into the space.

If you are struggling with the levers we have pressed, if you are questioning our explanations of MOO architexture and assembly, and if you are looking for hooks on which to hang our terms *cypher/TEXT* and *élekcriture,* we trust we have engaged you. We hope that you are prepared now to explore the following chapters in order to participate in their shifting boundaries, slippery non/surfaces, and multilayered folds and horizons—and to join in the reinvention of the *educational* MOO as a *cyphertextual* bricolage of scattered and cascading moments of *élekcriture.*

Notes

1. See the *High Wired* site (http://lingua.utdallas.edu/highwired/); the *High Wired* enCore educational MOO core site (http://lingua.utdallas.edu/hw/encore.html), and the Lingua MOO site (http://lingua.utdallas.edu/).

I

Concepts and Contexts

Far away, nearby, in the buzz of language
hums the tightrope walker to keep in balance
voice's body and body's voice, the humming
heavy as water.

There and not, the tightrope threads the current.
There and not, the walker reads the tightrope.
There and not, the song treads over the walker,
out of the body,

into mid-air, into the roar of language
run amuck in the circus. Meanwhile, above,
blind, the tightrope walker is stepping sideways,
lighter than water . . .

—Brian Clements

1

Finding One's Own in Cyberspace

Amy Bruckman

The week the last Internet porn scandal broke, my phone did not stop ringing: "Are women comfortable on the Net?" "Should women use gender-neutral names on the Net?" "Are women harassed on the Net?" Reporters called from all over the country with basically the same question. I told them all: your question is ill formed. "The Net" is not one thing. It is like asking: "Are women comfortable in bars?" That's a silly question. Which woman? Which bar?

The summer I was eighteen, I was the computer counselor at a summer camp. After the campers were asleep, the counselors were allowed out and would go barhopping. First everyone would go to Maria's, an Italian restaurant with red-and-white checked tablecloths. Maria welcomed everyone from behind the bar, greeting regular customers by name. She always brought us free garlic bread. Next we would go to the Sandpiper, a disco with good dance music. The Sandpiper seemed excitingly adult—it was a little scary at first, but then I loved it. Next, we went to the Sportsman, a leather motorcycle bar that I found absolutely terrifying. Huge, bearded men bulging out of their leather vests and pants leered at me. I hid in the corner and tried not to make eye contact with anyone, hoping my friends would get tired soon and give me a ride back to camp.

Each of these bars was a community, and some were more comfortable for me than others. The Net is made up of hundreds of thousands of separate communities, each with its own special character. Not only is the Net a diverse place, but women are diverse as well—there were leather-clad women who loved the Sportsman, and plenty of women revel in the fiery rhetoric of Usenet's alt.flame. When people complain about being harassed on the Net, they have usually stumbled into the wrong online community. The question is not whether women are comfortable on the Net, but rather, what types of communities are possible? How can we create a range of communities so that everyone—men and women—can find a place that is comfortable for them?

If you are looking for a restaurant or bar, you can often tell without even going in: Is the sign engraved wood or flashing neon? Are there lots of cars parked out front? What sort of cars? (You can see all the Harleys in front of the Sportsman from a block away.) Look in the window: how are people dressed? We are accustomed to diversity in restaurants. People know that not all restaurants will please them, and they employ a variety of techniques to choose the right one.

It is harder to find a good virtual community than it is to find a good bar. The visual cues that let you spot the difference between Maria's and the Sportsman from across the street are largely missing. Instead, you have to lurk—enter the community and quietly explore for a while, getting the feel of the place. Although published guides exist, they are not always useful—most contain encyclopedic lists with little critical evaluation, and by the time they are published they are already out of date. Magazines like *NetGuide* and *Wired* are more current and more selective and therefore more useful, but their editorial bias may not fit your personal tastes.

Commonly available network-searching tools are also useful. The World Wide Web is filled with searching programs, indexes, and even indexes of indexes (metaindexes). Although browsing with these tools can be a pleasant diversion, it is not very efficient, and searches for particular pieces of information often end in frustration. If you keep an open mind, however, you may come across something good.

Shaping an Online Society

But what happens if, after exploring and asking around, you still can't find an online environment that suits you? Don't give up: start your own! This doesn't have to be a difficult task. Anyone can create a new newsgroup in Usenet's "alt" hierarchy or open a new chat room on America Online. Users of UNIX systems can easily start a mailing list. If you have a good idea but not enough technical skill or the right type of Net access, there are people around eager to help. The more interesting question is: How do you help a community to become what you hope for? Here, I can offer some hard-won advice.

In my research at the MIT Media Lab (working with Professor Mitchel Resnick), I design virtual communities. In October 1992, I founded a professional community for media researchers on the Internet called MediaMOO. Since then, as MediaMOO has grown to one thousand members from thirty-three countries, I have grappled with many of the issues that face anyone attempting to establish a virtual community. MediaMOO is a "multi-user dungeon," or MUD—a virtual world on the Internet with rooms, objects, and people from all around the world. Messages typed in by a user instantly appear on the screens of all other users who are currently in the same virtual "room." This real-time interaction distinguishes MUDs from Usenet newsgroups, where users can browse through messages created many hours or days before. The MUD's virtual world is built in text descriptions. MOO stands for MUD object-oriented, a kind of MUD software (created by Pavel Curtis of the Xerox Palo Alto Research Center and Stephen White, now at InContext Systems) that allows each user to write programs to define spaces and objects.

The first MUDs, developed in the late 1970s, were multiplayer fantasy games of the Dungeons and Dragons variety. In 1989, a graduate student at Carnegie-Mellon University named James Aspnes decided to see what would happen if you took away the monsters and the magic swords and instead let people extend the virtual world. Players' main activity went from trying to conquer the virtual world to trying to build it, collaboratively.

Most MUDs are populated by undergraduates who should be doing their homework. I thought it would be interesting instead to bring together a group of people with a shared intellectual interest: the study of media. Ideally, MediaMOO should be like an endless reception for a conference on media studies. But given the origin of MUDs as violent games, giving one an intellectual and professional atmosphere was a tall order. How do you guide a MOO's evolution, who uses the space and what they do there?

A founder/designer can not control what the community ultimately becomes—much of that is up to the users—but can help shape it. The personality of the community's founder can have a great influence on what sort of place it becomes. Part of what made Maria's so comfortable was Maria herself. She radiated a warmth that made me feel at home.

Similarly, one of the most female-friendly electronic communities I have visited is New York City's ECHO (East Coast Hang Out) bulletin board, run by Stacy Horn. Smart, stylish, and deliberately outrageous, Horn is role model and patron saint for the ECHO-ites. Her outspoken but sensitive personality infuses the community and sends a message to women that it is all right to speak up. She added a conference to ECHO called WIT (women in telecommunications), which one user describes as "a warm, supportive, women-only, private conference where women's thoughts, experiences, wisdom, joys, and despairs are shared." But Horn also added a conference called BITCH, which the ECHO-ite calls "WIT in black-leather jackets. All-women, riotous and raunchy."

Horn's high-energy, very New York brand of intelligence establishes the kind of place ECHO is and influences how everyone there behaves. When ECHO was first established, Horn and a small group of her close friends were the most active people on the system. "That set the emotional tone, the traditional style of posting, the unwritten rules about what it's OK to say," says Marisa Bowe, an ECHO administrator for many years. "Even though Stacy is too busy these days to post very much, the tone established in the early days continues," says Bowe, who is now editor of the online magazine *Word*.

Beyond the sheer force of a founder's personality, a community establishes a particular character with a variety of choices on how to operate. One example is to set a policy on whether to allow participants to remain anonymous. Initially, I decided that members of MediaMOO should be allowed to choose: they could identify themselves with their real names and email addresses, or remain anonymous. Others questioned whether there was a role for anonymity in a professional community. As time went on, I realized they were right. People on MediaMOO are supposed to be networking, hoping someone will look up who they really are and where they work. Members who are not willing to share their personal and professional identities are less likely to engage in serious discussion about their work and consequently about media in general. Furthermore, comments from an anonymous entity are less valuable because they are unsituated—"I believe X" is less meaningful to a listener than "I am a librarian with eight years of experience who lives in a small town in Georgia, and I believe X." In theory, anonymous participants could describe their professional experiences and place their comments in that context; in practice it tends not to happen that way. After six months, I proposed that we change the policy to require that all new members be identified. Despite the protests of a few vocal opponents, most people thought that this was a good idea, and the change was made.

Each community needs to have its own policy on anonymity. There's room for diversity here too: some communities can be all-anonymous, some all-identified, and some can leave that decision up to each individual. (An aside: right now on the Net no one is either really anonymous or really identified. It is easy to fake an identity; it is also possible to use either technical or legal tools to peer behind someone else's veil of anonymity. This ambiguous state of affairs is not necessarily unfortunate: it's nice to know that a fake identity that provides a modicum of privacy is easy to construct, but that in extreme cases such people can be tracked down.)

Finding Birds of a Feather

Another important design decision is admissions policy. Most places on the Net have a strong pluralistic flavor, and the idea that some people might be excluded from a community ruffles a lot of feathers. But exclusivity is a fact of life. MIT would not be MIT if everyone who wanted to come was admitted. Imagine if companies had to give jobs to everyone who applied! Virtual communities, social clubs, universities, and corporations are all groups of people brought together for a purpose. Achieving that purpose often requires that there be some way to determine who can join the community.

A key decision I made for MediaMOO was to allow entry only to people doing some sort of "media research." I try to be loose on the definition of *media*—writing teachers, computer network administrators, and librarians are all working with forms of media—but strict on the definition of *research*. At first, this policy made me uncomfortable. I would nervously tell people, "It's mostly a self-selection process. We hardly reject anyone at all!" Over time, I've become more comfortable with this restriction and have enforced the requirements more stringently. I now believe my initial unease was naive.

Even if an online community decides to admit all comers, it does not have to let all contributors say anything they want. The existence of a moderator to filter postings often makes for more focused and civil discussion. Consider Usenet's two principal newsgroups dealing with feminism—alt.feminism and soc.feminism. In alt.feminism, anyone can post whatever they want. Messages in this group are filled with the angry words of angry people; more insults than ideas are exchanged. (Titles of messages found there on a randomly selected day included "Women & the workplace (it doesn't work)" and "What is a feminazi?") The topic may nominally be feminism, but the discussion itself is not feminist in nature.

The huge volume of postings (more than two hundred per day, on average) shows that many people enjoy writing such tirades. But if I wanted to discuss some aspect of feminism, alt.feminism would be the last place I would go. Its sister group, soc.feminism, is moderated—volunteers read messages submitted to the group and post only those that pass muster. Moderators adhere to soc.feminism's lengthy charter, which explains the criteria for acceptable postings—forbidding ad hominem attacks, for instance.

Moderation of a newsgroup, like restricting admission to a MUD, grants certain individuals within a community power over others. If only one group could exist, I would have to choose the uncensored alt.feminism to the moderated soc.feminism. Similarly, if MediaMOO were the only virtual community or MIT the only university, I would argue that they should be open to all. However, there are thousands of universities, and the Net contains hundreds of thousands of virtual communities, with varying criteria for acceptable conduct. That leaves room for diversity: some communities can be moderated, others unmoderated. Some can be open to all, some can restrict admissions.

The way a community is publicized—or not publicized—also influences its character. Selective advertising can help a community achieve a desired ambiance. In starting up MediaMOO, for example, we posted the original announcement to mailing lists for different aspects of media studies—not to the general-purpose groups for discussing MUDs on Usenet. MediaMOO is now rarely if ever deliberately advertised. The group has opted not to be listed in the public, published list of MUDs on the Internet. Members are asked to mention MediaMOO to other groups only if the majority of members of that group would probably be eligible to join MediaMOO.

New members are attracted by word of mouth among media researchers. To bring in an influx of new members, MediaMOO typically "advertises" by organizing an online discussion or symposium on some aspect of media studies. Announcing a discussion group on such topics as the techniques for studying behavior in a virtual community or strategies for using computers to teach writing attracts the right sort of people to the community and sets a tone for the kinds of discussion that take place there. That is much more effective than a more general announcement of MediaMOO and its purpose.

In an ideal world, virtual communities would acquire new members entirely by self-selection: people would enter an electronic neighborhood only if it focused on something they cared about. In most cases, this process works well. For example, one Usenet group that I sometimes read—sci.aquaria—attracts people who are really interested in discussing tropical fishkeeping. But self-selection is not always sufficient. For example, the challenge of making MediaMOO's culture different from prevailing MUD culture made self-selection inadequate. Lots of undergraduates with no particular focus to their interests want to join MediaMOO. To preserve MediaMOO's character as a place for serious scholarly discussions, I usually reject these applications. Besides, almost all of the hundreds of other MUDs out there place no restrictions on who can join. MediaMOO is one of the few that is different.

Emotionally and politically charged subject matter, such as feminism, makes it essential for members of a community to have a shared understanding of the community's purpose. People who are interested in freshwater and saltwater tanks can coexist peacefully in parallel conversations on sci.aquaria. However, on alt.feminism, people who want to explore the implications of feminist theory and those who want to question its basic premises do not get along quite so well. Self-selection alone is not adequate for bringing together a group to discuss a hot topic. People with radically differing views may wander in innocently, or barge in deliberately—disrupting the conversation through ignorance or malice.

Such gate-crashing tends to occur more frequently as the community grows in size. For example, some participants in the Usenet group alt.tasteless decided to post a series of grotesque messages to the thriving group rec.pets.cats, including recipes for how to cook cat. A smaller, low-profile group may also be randomly harassed, but that is less likely to happen.

In the offline world, membership in many social organizations is open only to those who are willing and able to pay the dues. While it may rankle an American pluralistic sensibility, the use of wealth as a social filter has the advantages of simplicity and objectivity: no one's personal judgment plays a role in deciding who is to be admitted. And imposing a small financial hurdle to online participation may do more good than harm. Token fees discourage the random and pointless postings that dilute the value of many newsgroups. One of the first community networks, Community Memory in Berkeley, California, found that charging a mere twenty-five cents to post a message significantly raised the level of discourse, eliminating many trivial or rude messages.

Still, as the fee for participation rises above a token level, this method has obvious moral problems for a society committed to equal opportunity. In instituting any kind of exclusionary policy, the founder of a virtual community should first test the key assumption that alternative, nonexclusionary communities really do exist. If they do not, then less restrictive admissions policies may be warranted.

Building on Diversity

Anonymity policy, admissions requirements, and advertising strategy all contribute to a virtual community's character. Without such methods of distinguishing one online hangout from another, all would tend to sink to the least common denominator of discourse—the equivalent of every restaurant in a town degenerating into a dive. We need better techniques to help members of communities develop shared expectations about the nature of the community, and to communicate those expectations to potential new members. This will make it easier for people to find their own right communities.

Just as the surest way to find a good restaurant is to exchange tips with friends, word of mouth is usually the best way to find out about virtual communities that might suit your tastes and interests. The best published guides for restaurants compile comments and ratings from a large group of patrons, rather than relying on the judgment of any one expert. Approaches like this are being explored on the Net. Yezdi Lashkari, cofounder of Agents Inc., designed a system called Webhound that recommends items of interest on the World Wide Web. To use Webhound, you enter into the system a list of web sites you like. It matches you with people of similar interests, and then recommends other sites that they like. Not only do these ratings come from an aggregate of many opinions, but they also are matched to your personal preferences.

Webhound recommends just World Wide Web pages, but the same basic approach could help people find a variety of communities, products, and services that are likely to match their tastes. For example, Webhound grew out of the Helpful Online Music Recommendation Service (HOMR), which recommends musical artists. A subscriber to this service—recently renamed Firefly—first rates a few dozen musical groups on a scale from "the best" to "pass the earplugs"; Firefly searches its database for people who have similar tastes and uses their list of favorites to recommend other artists that might appeal to the subscriber. The same technique could recommend Usenet newsgroups, mailing lists, or other information sources. Tell it that you like to read the Usenet group rec.arts.startrek.info, and it might recommend alt.tv.babylon-5—people who like one tend to like the other. While no such tool yet exists for Usenet, the concept would be straightforward to implement.

Written statements of purpose and codes of conduct can help communities stay focused and appropriate. MediaMOO's stated purpose, for example, helps set its character as an arena for scholarly discussion. But explicit rules and mission statements can go only so far. Elegant restaurants do not put signs on the door saying "no feet on tables" and fast-food restaurants do not post signs saying "feet on tables allowed." Subtle cues within the environment indicate how one is expected to behave. Similarly, we should design regions in cyberspace so that people implicitly sense what is expected and what is appropriate. In this respect, designers of virtual communities can learn a great deal from architects.

Vitruvius, a Roman architect from the first century B.C., established the basic principle of architecture as commodity (appropriate function), firmness (structural stability), and delight. These principles translate into the online world, as William Mitchell, dean of MIT's School of Architecture and Planning, points out in his book *City of Bits: Space, Place, and the Infobahn:*

> Architects of the twenty-first century will still shape, arrange and connect spaces (both real and virtual) to satisfy human needs. They will still care about the qualities of visual and ambient environments. They will still seek commodity, firmness, and delight. But commodity will be as much a matter of software functions and interface design as it is of floor plans and construction materials. Firmness will entail not

> only the physical integrity of structural systems, but also the logical integrity of computer systems. And delight? Delight will have unimagined new dimensions.

Marcos Novak of the University of Texas at Austin is exploring some of those "unimagined dimensions" with his notion of a "liquid architecture" for cyberspace, free from the constraints of physical space and building materials. But work of this kind on the merging of architecture and software design is regrettably rare; if virtual communities are buildings, then right now we are living in the equivalent of thatched huts. If the structure keeps out the rain—that is, if the software works at all—people are happy.

More important than the use of any of these particular techniques, however, is applying an architect's design sensibility to this new medium. Many of the traditional tools and techniques of architects, such as lighting and texture, will translate into the design of virtual environments. Depending on choice of background color and texture, type styles, and special fade-in effects, for instance, a web page can feel playful or gloomy, futuristic or old-fashioned, serious or fun, grown-up or child-centered. The language of the welcoming screen, too, conveys a sense of the community's purpose and character. An opening screen thick with the jargon of specialists in, say, genetic engineering, might alert dilettantes that the community is for serious biologists.

As the Net expands, its ranks will fill with novices—some of whom, inevitably, will wander into less desirable parts of cybertown. It is important for such explorers to appreciate the Net's diversity—to realize, for example, that the newsgroup alt.feminism does not constitute the Internet's sole contribution to feminist debate. Alternatives exist.

I am glad there are places on the Net where I am not comfortable. The world would be a boring place if it invariably suited any one person's taste. The great promise of the Net is diversity. That is something we need to cultivate and cherish. Unfortunately, there are not yet enough good alternatives—too much of the Net is like the Sportsman and too little of it is like Maria's. Furthermore, not enough people are aware that communities can have such different characters.

People who accidentally find themselves in the Sportsman, that is, alt.feminism or alt.flame, and do not find the black leather or fiery insults to their liking should neither complain about it nor waste their time there—they should search for a more suitable community. If you have stumbled into the wrong town, get back on the bus. But if you've been a long time resident and find the community changing for the worse—that's different. Don't shy away from taking political action within that community to protect your investment of time: speak up, propose solutions, and build a coalition of others who feel the same way you do.

With the explosion of interest in networking, people are moving from being recipients of information to creators, from passive subscribers to active participants and leaders. Newcomers to the Net who are put off by harassment, pornography, and just plain bad manners should stop whining about the places they find unsuitable and turn their energies in a more constructive direction: help make people aware of the variety of alternatives that exist, and work to build communities that suit their interests and values.

Note

This essay is reprinted with permission from the author. It appeared in *Technology Review* (January 1996) and is available online at http://web.mit.edu/techreview/www/articles/jan96/Bruckman.html.

2

Not Just a Game

How LambdaMOO Came to Exist and What It Did to Get Back at Me

Pavel Curtis

In the late summer of 1990, I was a researcher in programming-language design and implementation in the Computer Science Laboratory of the Xerox Palo Alto Research Center; I worked on compilers, formal semantics, and type theory. One day, I stumbled across the world of MUDs, and, in relatively short order, I started hacking on a particularly user-programmable MUD server called MOO. I was interested in MUDs almost entirely from the point of view of programming-language design; they struck me as an interesting context for this, with a number of unique challenges. It was a technical thing.

By the end of 1990, I was archwizard of LambdaMOO and had personally greeted almost every one of the hundreds of users there. By the fall of 1991, I was writing an amateur social-science paper about the experience of leading a large and growing online community. By the end of 1992, I had been threatened with lawsuits enough times to start laughing about it, I had a research group working on nonrecreational uses for MUDs, and my intellectual and personal lives had been dramatically transformed. Something big had happened, you understand.

This story is about a very personal impact of MOO: the transformation of a simple country hacker into a wizard, a judge, a community builder, and a wild-eyed visionary with delusions of anthropology. It is a story about the power of simple, mutable communications technologies to create living and breathing communities of people who care about each other and whose lives are deeply and permanently changed by the experience. It is also a story about how difficult it can be to operate such a system, to have responsibility for so volatile and powerful a space. Finally, it is a story of how one such place evolved and grew and changed, and of how these things will necessarily be different in the future.

```
New newsgroup alt.mud; subscribe? (yes/no)
```

I thought of mud wrestling, of course, or perhaps conversation among members of the Wet Dirt Appreciation Society. Given that this was Net-

News, and especially that it was the "alt." hierarchy, the possibilities were endless.

```
New newsgroup alt.mud; subscribe? (yes/no)
```

I paused. It's probably something really stupid, one of those newsgroups that gets created just because some weirdo out on the net wants some jollies, like "alt.meat.spam.more.more.more," one of those newsgroups whose name is itself the entirety of the content, itself supposed to be funny enough to warrant the effort of creating it.

```
New newsgroup alt.mud; subscribe? (yes/no)
```

I don't know, maybe you had to be there, but it didn't strike me as very funny.

```
New newsgroup alt.mud; subscribe? (yes/no) yes
```

Why not, I thought; maybe it would be an interesting puzzle, trying to figure out what they were talking about . . .

A Computer Scientist Trips and Falls into the MUD

I came to the Xerox Palo Alto Research Center (PARC) in the summer of 1983, as a summer intern working on adding static type checking to the Smalltalk programming language. That turned into my Ph.D. thesis work, on "constrained quantification in polymorphic type analysis." Along the way, I worked on several projects related to programming-language design and semantics, programming environments, progamming-language compilers and interpreters, and on and on. If it had to do with programming languages, I was interested in it.

By the time the summer of 1990 rolled along, I had started and was leading the SchemeXerox project in the Computer Science Laboratory at PARC; the project had been going on for over a year and was progressing quite nicely, I thought. At that time, though, CSL was going through a long process of introspection, reexamining many of its assumptions, in particular its priorities in the area of programming languages and environments. One distinct possibility was that the SchemeXerox project, among others, would be shut down entirely. Needless to say, these were not the most motivating circumstances; work on SchemeXerox slowed to a crawl. Discouraged, I tended to spend a lot more time reading NetNews than I had before; I was even willing to consider reading completely off-the-wall new groups whose charter was obscure at best.

```
New newsgroup alt.mud; subscribe? (yes/no) yes
```

As I read through the first day's traffic on this new group, a few things quickly became clear:

1. The discussion was neither about wrestling nor wet dirt.

2. Everyone who was contributing to the group already understood the subject matter so well that they saw no point in making it clear to anyone else.

3. Whatever this MUD stuff was, approximately half the contributors thought it was a game; the other half vehemently and heatedly disagreed.

I was not sure what to make of it, but I was sufficiently intrigued to want to stick around; the puzzle of it had me captured, and I wanted to keep reading at least long enough to figure out just what a MUD was.

Over the next few days, I ascertained that MUDs were some kind of software, that people used them together somehow, and that they owed something to older programs like Adventure and Zork. Now, Zork held a position of some honor in my memory; I had spent many of my first days on the old ARPAnet, logging into a machine at MIT every evening from my job in Berkeley, exploring and mapping and loving every part of the Great Underground Empire. Remembering the purely textual but evocative puzzles of Zork, clearly the product of wicked, twisted imaginations that I admired very highly, I became even more interested in tracking down these descendent MUD programs. It took, I think, over a week before someone on the newsgroup finally let slip enough information for me to connect to one of these worlds. I do not remember which one it was, but I do remember wandering around it for several hours, finding my way through a maze, and slowly coming to the realization that the MUD was not just another Zork, but one with a couple of very odd differences.

First, there were other people using the program at the same time, and it was possible for them to talk to one another! It is a little embarrassing to admit it now, but it actually took me some time to figure this out. As it happened, I had entered the MUD at a time when nobody else was active in the area of the entrance. Since I was just expecting something like Zork, I just read descriptions of things, found various exits, and just struck off on my own, completely oblivious to the "presence" of other people. You can easily imagine my surprise the first time someone "teleported" into the same "room" and started talking with me; I almost embarrassed myself further by mistaking them for a cleverly programmed robot!

The other thing that caught my attention came up in typed conversation with the first MUD player I met there, when he proudly took credit for having created the very maze through which I was laboriously moving. This program, this place, this virtual world was created by the very same people who were currently visiting it, along with a great number of others. *Anybody* who came there and explored for long enough was allowed to create more!

This point engaged me primarily because I wanted to know how this creation worked. What kinds of options did one have? How did one express the kind of behavior one wanted from the objects and rooms one created? I recognized immediately that, in some form at least, there must be a user-accessible programming language here. Being a professionally trained programming-language designer, I have a sometimes morbid fascination with the design efforts of amateurs; I wanted to see if, maybe, these folks had done something interesting (or at least passable) with this particular design problem.

After disconnecting from the MUD, I started tracking down and exploring the FTP sites that contained documentation and source code for the many kinds of MUD servers that existed. (I know it seems hard to believe now, but at that time there was no World Wide Web, no Yahoo or Alta Vista; the search was more difficult but, in some ways, also more rewarding then. Trust me on this.) As I searched and dissected these sites, I discovered to my disappointment that MUDs were, at least in the main, not really programmable at all; there were various commands for building and for setting the messages on the objects, but pretty much all of the actual behavior was fixed and unchangeable by the users.

After much searching, I finally came across two exceptions to this rule, two MUD programs that were almost entirely programmable by the users, where almost none of the behavior was built in. The first of these was very disappointing from a designer's point of view, being very error prone and difficult to use reliably. The second one, though, was different; it was reasonably straightforward and reliable, and it had a simple but powerful object-oriented programming model. A student from Waterloo, one Stephen White, had created it; he called it "MOO."

From MOO to LambdaMOO in Four Easy Months

It was late in September 1990 when I found the "alpha test MOO" running on a machine named "belch" in Berkeley. Stephen (aka Ghondahrl aka ghond) had set up this site so that he and others could test out his creation as it was being written and refined. It looked to me like a few tons of folks had created virtual objects and places there, and it did not take long to find out about the commands for inspecting those objects and places in detail; in particular, I spent a long time poring over the MOO programs that gave these virtual things their behavior. I still have the file into which I captured all of these program listings and from which I learned my first lessons about the interesting little programming language that Stephen had designed.

It was not perfect, of course (what language is?), but he seemed to have made few of the mistakes made by most language-design amateurs. There were even features to his language that were original, not simply cadged from other places; moreover, those features "fit" well with the rest of the language. I was moderately impressed. I wrote a few simple MOO programs of my own, trying to re-create in MOO some of my favorite puzzles from Zork, and decided after a few days that it would be fun to work with Stephen on the future evolution of his language. I approached him about this (on the MOO, of course), and he was receptive; it turned out that he had largely stopped work on the system anyway, so he agreed to send me a copy of the server program so that I could move it forward along some lines that appealed to me and my own sense of the Right Stuff of programming-language design.

We corresponded for a while as I sent him updates on my progress and ideas for future work and he responded with ideas, suggestions, and encouragements of his own. I tackled larger pieces of the existing code, fixing bugs and rewriting with an eye to making future changes easier to add. By the end of October, I had added a number of new features, fixed a number of bugs, and gotten hungry to try out my new version of the program on others. For my earlier MUD visits, I had used the name Lambda for myself; it was a major keyword in the Scheme programming language that I had been working in for so long, and so it naturally sprang to mind when first I was prompted for a character name. I now decided to use LambdaMOO as the name for my revision of Stephen's MOO server, since it was Lambda's MOO.

During my explorations on the Berkeley MOO, I had met a number of other people, some of whom were to have a major impact on the structure and style of LambdaMOO. I remember, in particular, one evening's conversation early in my tenure there; I chatted for quite some time with three other folks, and it became clear after a while that two of them, named Gemba and Gary_Severn, must know each other outside of this electronic context. Since they were also clearly among the most skilled of the MOO programmers there, I had them pegged as graduate students at Berkeley, near the server machine. The conversation eventually turned, as so many do on MUDs, to where we all were IRL (in real life); I was shocked to discover that Gemba and Gary were actually connected from the Australian National University and that the third person was sitting at a terminal in Israel! It was only then that the geographic scope of MUD participation began to dawn on me.

The day before Halloween 1990, I sent email to Tim Allen (aka Gemba), announcing that LambdaMOO was open to the world. I encouraged him to pass this information on to Gary_Severn, and on Halloween we three met there together, the first drops of water in what later became the rushing river of LambdaMOO.

During the earliest days of LambdaMOO, through the beginning of 1991, everything was fascinating every day. The technical work was especially so as Gary, Gemba, and I tried to build the core libraries of MOO programming, as I furiously wrote new functionality into the server and then furiously wrote new MOO commands that took advantage of it. I pounded the keyboard for hours at a stretch trying to write a comprehensive reference manual for this language and system.

The collaborative feel of it was also fascinating as we worked closely together from our separate offices thousands of miles apart, as we had good-natured arguments about the relative merits of American and Australian spellings of English words, as we were joined by Frand, a legendary programmer from the then-defunct Berkeley MOO; by Ford, a fellow researcher at PARC now thoroughly bitten by the MOO bug; and by yduJ, a hacker and former housemate we managed to attract away from some other MUD with the promise of a "real" programming language.

The creativity was fascinating as I laid out the core geography of the LambdaMOO mansion (based on the layout of my real-life house) and wrote descriptions of each room, as yduJ built the master bedroom suite, including LambdaMOO's first puzzle (the burglar alarm), and as Gary and Gemba followed their obsession into the design of the role-playing game infrastructure and its first few adventures.

But most fascinating of all was what was happening all around and through me: a community was forming inside this computer program, a community with hundreds of people, all learning about LambdaMOO by word of mouth and coming to see what it was about. Some of them were attracted by the programming language, as I had been, some simply by the newness and openness of the place, and some by something I had not planned or even really thought about at all: LambdaMOO was a *nice* place. On too many MUDs, the wizards were arbitrary, capricious, sometimes vicious; they played favorites, screwed around with the other, lesser players, and generally abused their privileged position in multiple immature ways. I did not come from the MUD culture, though; I had barely spent time on other MUDs, so I just treated folks more or less as I would have face to face: with respect and a naive, implicit expectation that everyone would behave well. Oddly enough, people picked up on that expectation and, in general, fulfilled it as if it were the most natural thing in the world; I guess, at the time, I was naive enough to think that it *was.*

Since we had not announced our presence to the world in any explicit fashion, the user community of LambdaMOO grew fairly slowly, one or two people at a time. I was connected to the place nearly all of the time I was awake, so I noticed whenever someone new connected and created a character for themselves. I used to personally welcome each and every new user for the first few months, at first just because I was hanging out with the crowd in the still-popular Living Room and later more intentionally, teleporting in from Lambda's Den, the private virtual office I had created for myself. It was neat, all these people from around the country and, as time went on, from around the world, coming to this program, this server, this place that I had created; they came, many of them stayed, and many of those helped make the place even better. It was magical. As what I later called a "simple country hacker," nothing like this had ever happened to me before.

Of course, not everyone was so nice or so constructive, though it took quite a while before this became clear. I think it was at least a few months into LambdaMOO's existence before the "Penn State assholes" came calling. I used to keep the server's network-connection log open on my screen all day while I was working, so that I'd know when new people connected. These two connections arrived nearly simultaneously, both from computers at Penn State University, and the log showed the player names these two new users had chosen for themselves: vulgar terms for parts of the female anatomy. Ah, I thought. A spot of trouble. A pair of lads (no doubt) who are a bit unclear on the concept. No problem, I thought; I'll just nip on down and straighten them out about what kind of place we've got here.

I caught up with them still in the coat closet, where all new players first connected. I pointed out that this was a place with lots of people around at once, where folks were connected from all manner of places, and where others might find the names they'd chosen to be, shall we say, somewhat objectionable. One of them replied with a belligerent question always guaranteed to annoy me: "What's your point?" I remained calm, asking him straight out to change his name before he went out into the living room. They both ignored me and went out anyway. By the time I joined them there, they were both typing "FUCK YOU" over and over again, to the annoyance of all present. I manually cut their network connections to the server, feeling both violated by their verbal attacks and depressed that I couldn't think of anything better to do. Later that day, in a black mood, I wrote the first version of the LambdaMOO '@toad' command, for permanently destroying a user's character and all of its possessions. Some of the shine was off the apple, never to return.

At PARC, we start each new year with a familiar ritual: the performance appraisal. Each researcher writes up a description of the work they have done in the past year, and their colleagues give written feedback on how it went. Around the beginning of 1991, I suddenly realized that I had spent the entire last quarter of the year on this "game" thing, this MUD! I had not done anything more at all on the SchemeXerox project I supposedly led, and now it was time to write about why. On a wild hope, I approached my area manager, pointed out that I had gotten a bit distracted during the latter part of the year, and asked if he thought it important that I write something about it in my PA. Oh yes, he said, he was looking forward to hearing more about this MUD stuff that had me so entranced. Sigh.

I wrote about MUDs primarily from a technical standpoint, bringing up some of the novel programming-language design issues that had captured and held my interest. I also mentioned that perhaps such systems could be used as a convenient testbed for ideas in computer support of collaboration. Looking back on that PA now, it seems very reasonable; at the time, I was distinctly uneasy about it. I remember a conversation I had in January 1991 with my lab manager, Mark Weiser, who had been aware of my growing interest in MUDs from the beginning (being himself a big fan of the old Zork and Adventure programs). In his indirect but nonetheless clearly understood fashion, Mark opined that, while this MUD stuff was amusing and perhaps even interesting on some level, I should be careful not to let it squeeze out the "real" research I was doing on SchemeXerox. Yes Mark, absolutely, message received, I'll do that. I had the very best of intentions, really I did.

LambdaMOO Faces the World and Vice Versa

By the beginning of February 1991, I finally felt ready to make a public announcement of the existence of LambdaMOO; we had a small but thriving community, enough structure in the virtual world to give newcomers a starting point, a reference manual worth reading, and enough online help texts to answer most questions we thought a new user would ask. I sent a message to the rec.games.mud newsgroup (alt.mud had gone mainstream in the meantime) saying where we were, where to find the documentation, and where to pick up a copy of the server software for creating new MOOs of one's own.

We got a fairly satisfying response, with lots of new users coming by to see what we had been up to. We stopped being impressed when there were ten users online at the same time and started more often seeing numbers like twenty-five and thirty. I thought we were huge beyond belief. With an increase in population and popularity, though, also came an increase in problems; we had built up a set of tacit rules for gracious living on LambdaMOO and had never really noticed ourselves doing so. These newcomers did not know our rules, did not know our style, and did not know the lessons we had learned over the course of four months of birthing. One persistent problem, for example, had to do with people discovering just how simple and general the MOO programming language was; it seemed for a long time that everyone's first MOO command just had to be the one that printed a message to every player who was logged on at the time, not just to the ones in the same virtual room. We had done this ourselves, some months before, and after the novelty had worn off, we had quickly decided that this "shouting" was almost never a good idea. Well, darn if these newcomers hadn't been there when we'd learned that!

Frand approached me one day to complain that several of his objects had been moved about by other players without his permission; he asked me to find a way to let all the new players know what was considered acceptable and unacceptable behavior on LambdaMOO. The idea seemed pretty daunting to me, but he had a good point, so I wandered all over the MOO asking old-timers (those who had been around since the good ol' days of October and November) for examples of unwritten rules that ought to get writ. After a few days, I wrote the first draft of the 'help manners' text that for a long time was the only written "law" that LambdaMOO had.

I was still writing a lot of MOO code at this time, and still making new public releases of the MOO server program, but I found that I was also spending a lot of time making the community run smoothly. I was often called upon to settle disputes between players, to interpret MOO law, and to deal with the occasional miscreant. When I was presented with the Case of the Lookalike Puppet, I realized just how much my daily routine was changing.

It seemed that one character and another had had a falling out; afterward, one of them had created a MOO puppet with the same name and description as the other player. (A puppet is a MOO object that acts much like a player, saying things, performing actions, etc., but is actually owned by and under the control of a player.) The owner of this puppet wasn't *doing* anything with it; it just sat in the same room as the owner, but with the name and description of the owner's former friend. The player being imitated was most upset, knowing that his doppelgänger was in Lambda-MOO. He wanted the puppet renamed at the least, and preferably outright destroyed.

It seems a bit silly and trivial now, but at the time I found it quite a puzzler. On the one hand, I could see how annoying it must be for the plaintiff, but on the other hand the puppet owner was not actively doing anything wrong. Since he never made the puppet speak or act, nobody was ever confused between the copy and the original; wasn't this a freedom-of-speech issue? I agonized over the situation, I had long talks with both players, I took the problem home to my wife and argued both sides of it with her.

I eventually worked out a compromise, but the case got me to thinking. What was going on here? What was happening to me? This was certainly not the kind of thing I had been used to doing as a programming-language researcher. I was a hacker and not a judge, wasn't I? Somehow, without my paying attention, my personal role was changing; somehow the clean boundaries between the technical and interpersonal worlds were shifting for me, changing and breaking down. This was pretty weird stuff, this MUDding.

I started telling stories to my friends about this weird place I had accidentally created, where all these seemingly normal people did such extraordinary things and behaved in such interesting and unusual ways. Since I personally found the whole thing fascinating to watch and participate in, and since it was all so new and different to most people, my stories got a pretty good reception. Eventually, toward the end of 1991, one of my friends told me about a conference coming up, run by Computer Professionals for Social Responsibility; the conference was called "Directions and Implications of Advanced Computing." My friend was on the organizing committee, and he thought it would be a good opportunity for me to write down my stories and tell a wider audience about my experiences.

I quickly agreed and almost immediately regretted it: I had no idea how to write this kind of a paper! I had written a variety of technical papers before, about programming-language design, compiler construction, and so on, so I knew something about writing, but I had never tried to write a scholarly paper about people instead of bits. I wandered the halls of PARC, at a loss for how to begin, and finally decided that it was time to consult an expert. What I was trying to do struck me as a form of anthropology; PARC, I knew, had a whole group of anthropologists as researchers, so I went off to the far end of the building to try to find one to help me. Hours later, I stumbled out of Francoise Brun-Cottan's office having just gone through nearly my entire stock of stories for her; at the beginning she had never heard of MUDs, but by the end she had helped me understand something about the axes along which social science is often organized. I wrote my paper in just about two weeks and submitted it to the conference committee. A few weeks later, they accepted it for inclusion in the conference and I began to see that there were more ways to do research on MUDs than I had previously considered.

In January 1992, I had another memorable conversation with my lab manager, almost exactly a year after the one I described above. He said that he realized that I had made some commitments to people concerning my SchemeXerox project, and he thought it important that I keep those commitments. He suggested, though, that after I'd done that I should consider going into "this MUD stuff" full time! It was clear to Mark, and becoming clear to me, that the work I was doing had a good chance of being important, and, more significantly, it looked like other computer scientists were not working in this area. That kind of opportunity does not come around for every young researcher; he thought, and I agreed, that I should seize it with both hands. What a difference a year can make.

LambdaMOO was first noticed in the popular press in 1992, and I will never forget how it began. On the morning of April 1, 1992, when I first got to work, I checked out the transcript of my perpetual connection to LambdaMOO. Amid the usual paged questions was a cryptic little message to the effect that a major fire had just swept through the house. Curious, I began wandering around the core of LambdaHouse. Everywhere I looked, the descriptions of the rooms had been changed to reflect ruinous fire damage; every little detail that I had put into the original descriptions was maintained in the new ones, but modified by the sad effects of the inferno. It was marvelous. Clearly, some of my staff of wizards had been very busy preparing for this wonderful April Fool's Day hack. I got a little worried that they had overwritten my original descriptions, but a little poking about revealed that the originals were safely preserved in hidden properties, waiting to be automatically restored at midnight.

At some point in my wanderings, a worried player paged me to say that it really, truly was not his fault, but he seemed suddenly to be a wizard! I didn't believe it, of course, but I checked just the same and discovered to my shock that it was true; when I inspected his player object, it clearly had the "wizard" bit on! His next page pointed to the latest article in the LambdaMOO newspaper; that article, written by my wizards, described the fire and said that, in order to hasten the repairs, all players had been made into wizards so that they could help out. I was utterly aghast. The fire was funny, but had they taken complete leave of their senses? Making everyone a wizard could only lead to chaos and destruction! I immediately typed the command to clear that first player's "wizard" bit and then checked to make sure it was, indeed, unset. But it wasn't! The inspection commands *still* showed that he was a wizard! Even through my panic, I now began to be suspicious; I listed out the code for the inspection command and, sure enough, found that a small change had been made: it would always show players to be wizards, even if they weren't. It was more of the April Fool's Day prank, and I was surely one of the fools; I stopped panicking and went back to being simply delighted with the joke.

Now, April Fool's Day jokes are an annual ritual in the high-tech firms in Silicon Valley; some companies, like Sun Microsystems, have such elaborate jokes played on their executives (by the employees) that they get local newspaper coverage each year. The LambdaMOO story got to be one of the "top ten jokes" in the next day's article, and the reporter came over later for a two-hour interview that led to the first full-length press story about LambdaMOO.

Things Fall Apart; the Center Cannot Hold

With the increased attention being paid to us, the population of LambdaMOO grew at an alarming pace. This, in turn, put an ever-increasing pressure on the wizards; we spent more and more of our time just keeping the place running, dealing with interplayer disputes and judging the things that players had built when they wanted permission to build more. None of us were being paid to operate LambdaMOO, but it was taking an increasing toll; we started to create institutions and automated procedures to lighten the load.

The first step was the institution of player registration. From the beginning of LambdaMOO, anyone could create a new character by typing the 'create' command at the login screen. This was now causing problems in resolving interplayer disputes, since nothing prevented a single human from racking up a large collection of different characters; if one of them got into trouble, the player could switch over to a new and different one and so evade responsibility. When one of the wizards volunteered to take on the chore of managing the process, we agreed to start registering players. Now, when a new character was created, it was necessary to provide a working email address for it; the address was validated by sending the randomly generated password for the new character only to that email address. In the short term, this actually added to our burden, as tools were written to automate the character-generation and password-mailing process, but the long-term effects were more beneficial, as players realized that there was now a level of true accountability on the MOO.

The other major drain on the wizards was the building quota system. I had put this in place at the beginning of LambdaMOO to keep the database from filling up with useless objects. Each player had a quota, the number of MOO objects they were allowed to create; only wizards could change that number and would only do so if the player was making good use of the existing quota. Making this judgment, though, is a time-consuming process; the wizard has to tour everything they have built, try out all of the commands, read all of the descriptions, and so forth. When one of my wizards threatened to quit if the workload did not decrease, I decided to move this quota-judgment task off all of our plates.

I created a committee of longtime LambdaMOO players to take over the judgment job, and, as a little joke, I named them after the kind of groups I had seen in RL (real-life) homeowners' associations: the Architecture Review Board (ARB). This group needed slightly privileged tools so that they could make a sufficiently complete judgment of other players' work. It seemed reasonable to create a special room to host those tools, so I made what I thought was another little joke, naming the new room the Star Chamber:

> star-cham-ber \'stär-'chām-ber\ *adj* [*Star Chamber,* a court existing in England from the 15th century until 1641] (1800): characterized by secrecy and often being irresponsibly arbitrary and oppressive

I admit, the joke is a little hard to see now; I was soon to learn that the administration of LambdaMOO was not the place for such jokes. From the standpoint of a random player on LambdaMOO, much about its operation is unclear; what might seem funny or outrageous or even absurd to one who is deeply entwined in the day-to-day events of the wizards can seem entirely believable, chilling, or demeaning to one outside the inner circle.

From the beginning of the ARB, some players were suspicious. How was it formed? Who chose the members and why? How do they make their decisions? What is said in the Star Chamber? Why can't we go in there? At first no one knew of anything bad actually happening around the ARB; its very existence and the way it was created were enough to worry some players. It is telling that one of the first ballot measures to pass when elections were put in place made ARB positions elective.

Another major burden on the wizards was interplayer disputes. The wizards were the police, the judges, and the executioners; we had set ourselves up for this back when "help manners" was drafted, in which we had claimed, "Vengeance is ours, sayeth the wizards." It *was* better that players appeal to the wizards than take revenge on each other, but there was a limit to how many players we could supervise. In retrospect, the last half of 1992 was characterized for the wizards by weariness and stress. It was *hard* trying to care for such a large and growing place, trying to care about the petty squabbles of so many people, trying to be everything to everyone. Something had to give, and it was us.

On December 9, 1992, I posted a pivotal message to LambdaMOO's *Social-Issues mailing list, titled "On to the next stage," but somehow history has tagged it "LambdaMOO Takes a New Direction," or "LTAND." I announced the abdication of the wizards from the "discipline/manners/arbitration business"; we would no longer be making what I glibly termed "social decisions." It was so simple in my mind: there was a clear distinction between social and technical, between policy and maintenance, between politics and engineering. We wizards would become technicians and leave the moral, political, and social rudder to "the people." In the metaphor of that message, I was throwing the baby bird of LambdaMOO society out of the nest of wizards as mothers; in hindsight, I forced a transition in LambdaMOO government, from wizardocracy to anarchy.

The transition was not immediately obvious to the average player on LambdaMOO (indeed, perhaps no such transition there has been universally felt; oddly enough, a lot of people go there simply to play around), but it had an instantaneous impact on the wizards. On the one hand, we felt great relief; an enormous burden had been lifted. Our lives were suddenly simpler; when someone complained to us about some other player, we no longer had to do anything! But there was also a dark side to the transition; it turned out that some players had been holding themselves in check only because they feared our wizardly response. One of the wizards was suddenly the target of harassment, verbal assault, and taunting: "What're you gonna do to us, eh? *Toad* us? Ha ha!" It was a sad day when the wizards of LambdaMOO decided they had no choice but to @gag certain other players, refusing to hear their taunts.

LambdaMOO became a rougher place after LTAND. It's hard to say how much LTAND accelerated a process that was already under way, but surely it did not help to hold it back. The level of interplayer strife and harassment rose, slowly but inexorably. The crisis came about four months after LTAND, in the infamous "rape in cyberspace" case.

I won't go into gory detail about that regrettable incident (Julian Dibbell has already covered it awfully well in his *Village Voice* article: ftp://ftp.parc.xerox.com/pub/MOO/papers/VillageVoice.txt), but suffice it to say that it was a particularly vicious piece of verbal sexual harrassment. The perpetrator never denied any of the accusations made against him, and there were widespread public demands for his toading. This created a quandary for the wizards: I had said in LTAND that the burden of such decisions now belonged to "society at large"; was this public outcry the expression of that society's will? Was it a consensus of the populace, one that the wizards, as the technicians of society, should implement? Eventually, hearing no opinions to the contrary, one of the wizards did the deed; the bad guy was toaded.

I was away on a business trip during the entire affair; I returned just in time to see the enormous backlash against the toading. Many players had not bothered to oppose the demands for toading, knowing that the wizards had *promised* not to take such social actions. Betrayed in that trust, they came out in full force, anger directed at the wizards; bitterness, acrimony, and belligerence reigned. It was clear, as one *Social-Issues poster said at the time, that with LTAND the wizards had abdicated responsibility without delegating it; the people were supposedly in charge, but no mechanism existed by which they could unequivocally state their intentions. In hasty response, one of the wizards tried to whip together a "petition" object: if a petition got one hundred signatures, the wizards would do whatever it said. I stepped in at this point and tried to fix my earlier mistake: I wrote my second pivotal message to *Social-Issues laying out what became the LambdaMOO petitions-and-ballots system. Once again, and again by fiat, I forced a governmental transition on the MOO: we would, I declared, move from anarchy to democracy.

Direct Democracy: LambdaMOO Is to Sysiphus as . . .

My primary inspiration for the design of the petitions-and-ballots system was the voter-sponsored initiative process in California. It was simple in outline: any LambdaMOO citizen could create a petition proposing that the wizards take some action; if it got enough signatures, it became a public ballot measure that passed on a two-thirds majority vote. Before a petition could become a ballot measure, though, the wizards had to "vet" it to ensure that the proposal was *(a)* clear, *(b)* feasible, *(c)* appropriate, *(d)* legal, and *(e)* secure. (The only one of these that is tricky to define is appropriateness; the idea was that the petition had to address some social goal and require wizardly action for its implementation.) The goal was to prevent ballots that the wizards could not implement without *(a)* making up an interpretation of the text (a social judgment), *(b)* getting into RL legal trouble, or *(c)* compromising the functional integrity of the server.

Almost from the very beginning, we got into trouble with vetting. It was much harder than I had anticipated to make sure, in advance, that a petition spelled out enough details to make its implementation a matter of pure engineering. The first petition to reach ballot status created LambdaMOO's controversial system of binding arbitration for interplayer disputes. The proposal was very sketchy and, in retrospect, should not have passed the vettins process. But we were new to this job and naive: we thought it was clear enough, that the details remaining to work out would not cause problems. Instead, the arbitration system has been plagued throughout its entire existence with ambiguity and allegations of corruption, wizardly bias, undue manipulation, and moral bankruptcy.

Of course, there were other ballots, even fairly major ones, whose vetting and implementation we got right and whose impact has been relatively noncontroversial. Examples include the introduction of byte-based quota (replacing the original object-based system), elections for the ARB, and the measures that reined in LambdaMOO's population growth.

Overall, though, the petition system has failed on LambdaMOO. It has failed to be the jumping off point I hoped for; we have not seen it used successfully to move LambdaMOO to a working, stable form of self-government. There were long periods, indeed, when many petitions reached ballot stage and failed to pass; it seems to me now that the voting population could never agree on any measures of real substance. This is the real lesson of LambdaMOO's experiment with direct democracy. In a representative democracy, politicians get together behind closed doors and make deals; they compromise in order to make progress. Direct democracy, though, leaves no room for this pliancy; direct democracy is intrinsically mob rule, and mobs are organizationally incapable of compromise. On LambdaMOO, this incapacity engendered a profound stagnation; true progress is impossible to achieve in the petition system.

Over three years have passed since the imposition of the petition system, years of crisis and response, of accusation and defensive reply, and sometimes of hope and joy. LambdaMOO has been terribly important to a number of people; I have the letters to prove it, from now-married couples who met there, from people who needed a friend and found one on the MOO, and from people who have learned a new self-esteem from the act of creating objects there and seeing other players' reactions to them. It is terribly important to me that I have those letters, that people have written to thank me for creating the place. These letters are important primarily because they are the exception, not the rule.

The structure of LambdaMOO's politics, this contrived theory of a clean separation between the technical and the social, mixes very poorly with the reality of the MOO's implementation. LambdaMOO runs on a single computer, and that computer is owned by some one entity; the underlying economics of the system are inherently unbalanced. The LambdaMOO database is a single, monolithic file, with every part tied to every other part by a complex chain of unbreakable dependencies; it simply is not possible for a group of players to effectively "secede" from LambdaMOO, to take their toys and go home. The server requires a nearly constant stream of little attentions from the wizards; it must be rebooted regularly, old unused accounts must be purged, bounced mail from new-account registrations must be dealt with, and the list goes on and on. Deep in its very structure, LambdaMOO depends on the wizards and on the owner of its machine. These are not and cannot be purely technical considerations. Social policy permeates nearly every aspect of LambdaMOO's operations, and only the wizards can carry out those operations.

As a result, the wizards have been at every turn forced to make social decisions. Every time we made one, it seemed, someone took offense, someone believed that we had done wrong, someone accused us of ulterior motives. It felt a bit like the laws of thermodynamics: you can't win, you can't even break even, and you can't get out of the game. The result was a nearly constant stream of messages to the wizards full of anger, suspicion, and of stress. Once again, in the spring of 1996, the wizards were close to the breaking point; some were leaving, some were taking vacations from the job, and everyone was feeling the strain. In one private communication after another, the wizards were telling me that something had to be done.

Throughout the entire month of April, the wizards' private mailing list was pulsing with activity; we were drafting, arguing about, and redrafting a new fiat, my third pivotal message to *Social-Issues. Finally, on May 16, we all agreed on a draft and I posted LTAD: LambdaMOO Takes Another Direction. In it, we formally repudiated my earlier theory of a social/technical dichotomy; we explained how impossible that fiction was and declared our intent to cease apologizing for our failures to make it reality. It was, in a way, a wizardly coup d'etat; out with the old order, in with the new.

Can We Get There from Here?

Will it work this time? Can we achieve peace on LambdaMOO and a tolerable job for the wizards? It's best not to set one's sights too high for a place like LambdaMOO. Trapped in a fundamentally untenable economic and political situation, perhaps it is enough to maintain, to keep the place together for another year or two or three, until it stops providing something that its thousands of regular users cannot get anywhere else, or until something better comes along.

What would "something better" look like? On the top of my personal list is a vastly more equitable distribution of both power and economics. If these online "virtual" communities are to have the same robustness as the more physically oriented ones, they must become pliant and mutable under the same forces that cause RL communities to grow and change. It must be possible for incompatible subcommunities to separate and grow apart, thereby relieving the kinds of stresses that constantly tear at LambdaMOO. It must be possible for individuals to completely control their own creations, without the specter of an all-powerful wizard looming in the background, distorting all natural social interactions.

When these things happen, when online communities can form and break apart as fluidly and naturally as do face-to-face ones, something dramatic will be taking place in the history of the human race. If you look closely at the Internet today, you can begin to see it happening already, in newsgroups and in the World Wide Web. It is still not here in the area of more synchronous interactions, but I know for a fact that it is coming; I am working hard for it every day. We have only to hold on, to keep dreaming the dream; more than at any other time in human history, dreaming is nearly enough to make it happen.

II

How to Use, Set Up, and Administer a MOO

. . . . endlessly falling
into the darkness, stopped from falling by reefs
of charge. There is a way to do this without
weight, the way a particle of lead hides

its double life as wave. And what we're doing
is work, the tightrope walker stepping sideways
to find the circus unthreaded, watched to death
like naked light. Where does light go when it stops?

Does the tightrope walker stop dead when he thinks
of the dual nature of the rope? Does he
question his religion? Or does it make things
easier to imagine himself falling?

Brian Clements

3

How to MOO without Making a Sound

A Guide to the Virtual Communities Known as MOOs

Jorge R. Barrios and Deanna Wilkes-Gibbs

Definition: MOO (mōō), intransitive verb: mooed mooing moos. The act of using a network-based virtual reality known as MOO (for Multi-User Domain/Dungeon, Object-Oriented), by which people across the world can interact with one another in real time.

Sample usage: "We mooed, we are mooing, we will moo . . . and maybe someday we'll even get some sleep!"

This chapter is intended to provide readers with the knowledge they will need to begin using a MOO productively, along with guides to sources of information that will also be of value to more experienced users. We begin with a basic introduction to using a MOO, including the minimal computing capabilities one needs as well as enhancements that can make using a MOO easier and more pleasurable. We then lead the reader step by step through the process of connecting to a MOO for the first time and explain how to become a permanent community member.

Throughout our discussion, we will assume neither technical expertise nor prior experience with MOOs. We will begin with just enough detail to get started and then unpack more information in subsequent sections. We will explain how to find help, how to talk and send mail to others on the MOO, and how to navigate the virtual space. One of the most distinctive features of MOOs is their extensibility. Users can literally change the world, and we will explain how to create new things and build new places. Finally, for those who wish to add more capabilities and tools to the MOO, we also provide a quick look at some useful MOO programming commands, as well as instructions on where to find more detailed technical information.

Although MOOs differ to some extent in the capabilities and commands they offer, the central functions are accomplished in a similar manner on all MOOs. The commands and capabilities we describe here should be available on all the commonly used versions of MOO software (at the time of this writing, the latest release was version 1.8.0p6 of the Lambda-

MOO server code). When a common point of reference is needed for purposes of explanation, we refer you for examples to the *High Wired* enCore educational MOO core database provided for use in conjunction with this book.

Any how-to guide needs consistent conventions of notation, and the conventions we adopt are the ones you will see in help texts within the MOO environment. In normal run of text (as here), we use single quotes to indicate material that should be typed verbatim. If you are asked to type '@who Sky', you should type the text between the single quotes as shown (but not the quotes themselves) and then press the return/enter key. Double quotes, in contrast to single ones, are generally meaningful in MOO commands and should be typed if they appear in an example of command syntax.

Angle brackets <like this> indicate that you should not type the text enclosed in the brackets verbatim but rather insert a value, such as an object number (#5), or a string ("Meridian Plaza"). Square brackets [like this] indicate that the parts of a command inside the brackets are optional. In both cases, you should not type the brackets themselves. They are just notational devices, not part of MOO command syntax.

In general, commands end in a return, which signals that you are ready to execute the command. We will assume this in most of our examples. Finally, MOO commands are case insensitive. Typing 'PAGE HELP' will have the same effect as typing 'page help' or 'Page Help' (although you should know that SAYING things in uppercase letters is usually interpreted as yelling). The only exception to this rule is your password, in which case does matter.

1.0 Before You Begin: Technical Prerequisites for Using a MOO

In this section, we describe the equipment and capabilities one must have to begin using a MOO, as well as some optional tools and enhancements that can make MOOing easier and more pleasurable.

1.1 The Bare Minimum

MOOing requires four essential ingredients. You must (1) have access to a computer, which in turn (2) has access to the Internet and (3) to a utility called Telnet for connecting to other machines over the network, and (4) you need to know the Internet address for a MOO. Each of these basic prerequisites is explained below. Finally, although not a requirement for using a MOO as a guest, you typically need a working email address of your own to become a registered member of a MOO community.

1.1.1 Your Computer

In the jargon of the technical world, a MOO is a *client-server application.* Basically, this means that the MOO runs on one computer (the server) but you interact with it by typing at the keyboard of your own computer or computer terminal (the client machine). You can use any kind of computer for MOOing (e.g., PC, Macintosh, workstation), so long as it has been properly set up for accessing the Internet.

1.1.2 Internet Access

Your computer needs a channel for communicating with the MOO server machine, and this channel is typically the Internet. How to find out if your computer is Internet-ready and how to get Internet access are subjects beyond the scope of this chapter. Fortunately, many educational institutions provide faculty, staff, and students with an Internet connection. Setting up an account with a private Internet service provider (for which you pay a fee) is getting easier every day. Note, however, that if you want to MOO from a computer at home, you will most likely need to dial in to your Internet access point (whether that be a machine at your school or work location, or a private Internet service provider). At minimum, this requires that your computer have a modem connected to your telephone line.

1.1.3 Telnet Capability

Telnet is a widely used utility program that connects one computer to another through the network and is the most common method of connecting to a MOO. There are versions of Telnet for many different types of computers and operating systems (e.g., Windows, Macintosh, UNIX, VMS), all of which accomplish the same function. Your institution's computing-services department or your Internet service provider will almost certainly be able to provide you with information about how to use Telnet on your computer.

1.1.4 A MOO'S Network Address

The Telnet program needs the address of the machine on which a MOO is running in order to let you connect to it. The Internet address of a MOO is the complete name or numerical identifier of its server machine, plus a number for the computer port through which the MOO is accepting connections. For example, the address of our education-related MOO, Meridian, is sky.bellcore.com, port 7777. In some cases, it might be necessary to use the numerical form of the address (commonly known as an *IP number*) instead of its name. In the Meridian example, that would look like: 207.3.224.160 (with the same port, 7777). Normally, however, the IP number and name are interchangeable.

Don't worry if none of this makes much sense at this point in your MOO endeavors. Just have a MOO's network address handy when you get ready to connect to it for the first time. The addresses for some education-related MOOs at the time of this writing are given in the appendix.

1.2 Useful Enhancements

Telnet is the means by which most new MOOers connect to a MOO, and many "regulars" still use only this method. However, another way to connect is by using a special "client program," installed directly on the computer you are using. Telnet itself is actually a sort of client program, but other more specialized clients have been developed for MOOs (or MUDs) to make your virtual life easier.

One of the main reasons for turning to a specialized client program is that Telnet was not developed with the highly interactive nature of MOO communication in mind. As a consequence, your own lines of typing can get mixed up on your screen with the output from a MOO (e.g., when someone there talks to you at the same time you are typing something you want to say to them). The specialized client programs can filter the output of Telnet to make it more readable on your screen. Most, for instance, separate your typed input from the MOO output, and do word wrapping so the MOO output text fits the size of your screen or window. Some also allow you to create shortcuts (or "macros") for frequently used MOO commands, and some provide tools that simplify building and programming in the MOO environment.

There are by now client programs for nearly every popular type of computer and operating system. Which one is best for you is largely a matter of the kinds of things you do on a MOO and your personal tastes. Some of the most popular clients are listed in table 1; these programs are either freeware or shareware and are available for you to download over the Internet from the web site that augments this book. If you do not have access to a web browser, these and other clients are also available for downloading by anonymous FTP to ftp.math.okstate.edu/pub/muds/clients. If you do not know how to download files from remote computers, ask your system administrator or Internet access provider.

MOOs have traditionally been plain-text environments, but several are now incorporating the World Wide Web's graphical and multimedia

TABLE 1. Examples of Some Client Programs for Connecting to MOOs

Operating System	Client Program
Macintosh	MUdDweller
OS/2	TinyFugue
Unix	TinyFugue
	Mud.el
	tkMOO
VMS	TinyFugue
Windows 3.11	Zmud
Windows 95	Gmud

capabilities. The ability to share graphics and to bring outside information into the MOO environment via the Web is an especially nice development for educational applications. To take advantage of the capabilities of these "webbed" MOOs, one needs World Wide Web access, a graphical web browser (such as Netscape or Internet Explorer), and the (HTTP) address for a MOO's web gateway. There are, in addition, special client programs under development specifically for use with "webbed" MOOs. At the time of this writing, two in early "beta" release were Cup-O MUD, written by Alex Stewart, and Surf_and_Turf, written by Ken Schweller (see chap. 5).

Describing use of the various webbed MOO interfaces now in existence is beyond the scope of this chapter, but you will find that all the text-oriented commands we describe here work in much the same way on a webbed MOO. We encourage you to explore this step in MOO evolution

2.0 Your First Time: The Bare Basics

So you have your Internet-connected computer, you know how to start Telnet or a specialized MOO client, and you are ready to experience a MOO firsthand. This section leads you step by step through the process of visiting a MOO for the first time, as a "guest." In the instructions we will presume that you are using only the Telnet utility to connect to the remote MOO computer. If you have followed our suggestion and obtained a special client program, you will follow the same steps, though you may accomplish some tasks more easily.

You can connect to a MOO as a guest by following the steps outlined below. The remainder of this section explains each step in greater detail.

> To reach the MOO, activate your Telnet program and give the command for it to create a connection to the address and port number of the MOO.

At the MOO's login screen, type 'connect guest' (and press the return/enter key). Follow any instructions you see on your screen during the connection process.

When you are finished exploring, type '@quit' to leave the MOO (and then close your Telnet connection if necessary).

2.1 Reaching a MOO

To reach a MOO over the Internet from your computer, you must give your Telnet program (or special client) the proper Internet address and port number for the MOO you wish to visit. So, for example, if you were using Telnet on a UNIX system and wanted to connect to Meridian MOO, you could type the following (at your UNIX prompt): 'telnet sky.bellcore.com 7777'.

2.2 The Login Screen and Welcome Message

You will know you have successfully reached the entryway to a MOO when you see its welcome message. The specific text you will see on your screen differs from one MOO to another, but at minimum you will see the name of the MOO (preferably the one you intended to visit!), and what your command options are at this point. To enter the MOO, you must now choose one of the options to "connect". In the next section we describe how to connect as a guest—that is, someone who is not already a member of the MOO.

2.3 Connecting as a Guest

Assuming that you do not already have your own character or persona on the MOO you want to visit, the way to enter it is by connecting as a guest. Although some MOOs allow anyone to create a permanent character "on the spot" by entering a command at the welcome screen, most education-related MOOs do not. Instead, one first visits these MOOs by using a guest character that allows you to explore the MOO and communicate with others there, but not to build or create things that change the MOO.

To connect as a guest from the welcome screen, type 'connect guest' and press return/enter.

2.4 Customizing Your Guest Character

On some MOOs you are given the opportunity to choose a name for your guest character (that name will be erased when you leave the MOO). On others, you are identified by one of their standard guest names, such as *Red_Guest*. Another option you might have is to give the guest character a description that people will see when they look at you (within the virtual reality, of course), and to choose a gender for your character.

2.4.1 @Describe

You can describe yourself (and other things you own, once you have a nonguest character) by using the @describe command. The command is

```
@describe me as "<the text of the description>"
```

For example, imagine that Red_Guest types '@describe me as "A tall man wearing glasses and a crumpled fedora."' Now this is the text that anyone sees if they type 'look Red_Guest'.

You can see how your description looks to others by typing 'look me'. One thing to note is that some MOOs show the Internet sites from which a guest is connecting to the MOO appended to his or her description. This is done to encourage guests to feel accountable for their behavior because some have used guest characters to flout norms of mannerly social conduct. Of course, MOO administrators always have access to both guest and regular characters' connect sites (as well as a variety of other information, potentially), but in the case of regular characters this information is typically treated as confidential.

2.4.2 @Gender

You may wish to tell the MOO what gender to apply to your character. Among other effects, this determines the pronouns that refer to you in MOO texts. Typical options are male, female, and neuter; some MOOs offer more unusual ones as well (type 'help @gender' while on the MOO to see your choices). To set your gender, type '@gender <your choice>': For example, '@gender male'.

The commands @describe and @gender are identical when you get your own, more permanent character on a MOO. (A minor note: some longtime MOO users consider it unmannerly to use either a guest or a regular character without setting a description and gender.)

2.5 What to Do First

As a guest, you "emerge" in a special area of the MOO where guests and other characters who do not have their own "homes" begin. This is frequently a "quiet" area (though there are exceptions) and a good place to begin learning about the MOO you are visiting. Typing 'help manners' is always a good way to learn about conventions of behavior. Typing 'help theme' should tell you about the purpose of the MOO, and you are also likely to find a variety of useful information in the MOO's newspaper (type 'help news'). Finally, typing '@who' will show you who is also connected to the MOO at this time, along with their virtual locations in it.

If the lines of MOO output are being cut off at the edge of your screen, you can get the MOO to split lines for you. The @linelength command tells the MOO how many columns you have on your screen; typing '@line-length 79' sets lines up for a screen width of 79 characters (a common default). The '@wrap on' command then tells the MOO you want it to do word wrapping according to the line length you've specified. You can type 'help @linelength' for more detailed information.

2.6 Communicating with Others

MOOs are usually friendly places, and people often assume that others on the MOO are open to interaction unless there are indications to the contrary. As a result, one of the first things likely to happen is that someone will initiate a conversation with you. In this section, we summarize the most basic forms of MOO communication, 'say' and 'emote', so that you will be able to respond. Look for more details and nuances in section 4.0.

2.6.1 Say

To speak to someone in the same virtual room, use the 'say' command. Simply prepend the word 'say' to whatever it is you wish to say. For example:

```
Knekkebjoern types: say Hi, Sky. How are you?
Everyone in the same room sees: Knekkebjoern says, "Hi,
  Sky. How are you?"
```

The say command can be abbreviated as a quotation mark (").

2.6.2 Emote

To appear to do something rather than say it, use the emote command. Note that the text of an emote typically begins with a verb, as exemplified below:

```
Knekkebjoern types: emote yawns and stretches.
Everyone in the same room sees: Knekkebjoern yawns and
  stretches.
```

The emote command can be abbreviated by a colon (:).

2.6.3 Page

To speak to someone who is not in the same MOO room, use the page command. The syntax is 'page <character> <message>', as in the following example:

```
Sky types: page Knekkebjoern Long time no see!
Knekkebjoern sees: You sense that Sky is looking for
  you in Sky's Office. She pages, "Long time no see!"
```

Each character (like all other objects) has an object number that can be used in place of the character name in commands. If Knekkebjoern's object number is 199, 'page #199 Long time no see!' gives the results shown in the preceding example.

2.7 Navigating the MOO Space

Many MOOs have virtual subways, trains, airports, or even hot-air balloons, and they may allow you to tour a MOO's highlights. Some MOOs also offer organized tours, with either live or robot guides, and these can be especially useful for newcomers. Such tours are usually advertised close to the point of entry for guests.

2.7.1 Walking

Should you not choose a tour, there are several ways to navigate the virtual spaces of a MOO on your own. The simplest is to "walk" from one room to another by typing the name of the exit that connects them. Exits are commonly shown in a room's description, which you see when you enter a room for the first time or by using the 'look' command. If you forget where you are at any given time, you can always look at the room again or type '@who' to see where you and all the other people on the MOO are located at any given moment. Typing 'home' will return you to your starting point, should you get lost or stuck somewhere.

2.7.2 @Go

On many MOOs there are efficient means of traveling to a specific destination. One of these is the '@go' command. If you know the object number of a room you would like to go to, you can get there quickly by typing '@go <#room number>'. On some MOOs you may be able to use the room's name in the @go command instead of its number.

2.7.3 @Join

Another very useful command is '@join'. If you type '@join <character>' or '@join <#character's object number>', you will magically be transported to the room in which that person is currently located (providing that it is not locked; see section 7.0). The person does not have to be connected in order for this to work; that is, you can @join a sleeping (i.e., unembodied) character. It is worth noting that conventions differ as to whether it is considered impolite to @join someone without first having gotten permission (e.g., by paging). On some MOOs, for example, one is expected first to use the command '@knock <character>' and wait for an invitation before joining them. On others, however, it is perfectly acceptable to drop in on someone unannounced. When in doubt, err on the conservative side by using @knock or paging to ask if it's OK to join.

2.8 Getting Help on the MOO

2.8.1 The Help Database

For routine questions, how to do or find something on a MOO, the first place to look is the in-MOO help system. One can query the help database by typing 'help <word>'. Many, many operations are documented in the online system. Often the problem, however, is knowing what word to type as the query. You can get started by typing 'help help' or 'help full-index'.

2.8.2 Page Help

If your attempts to get help from the online database fail, the next source is a more experienced person who is connected at the same time you are. In fact, many MOOs have a cadre of helpful people who have volunteered to respond to pages for help. Try typing 'page help' to see if the MOO you are on has such help. If it has, your page will be directed to one of the helpers who is connected, and you should receive a response.

2.8.3 In Case of Emergency

If you find yourself in an emergency, type '@wizards' to see what administrators are connected and then page one of them for help. (Remember, you page someone by typing 'page <character> <message>' and pressing return/enter.) You can also page help or page anyone else if no wizards are connected. Most MOOers will gladly offer assistance to someone in need.

For nonemergency questions, you should generally page an administrator only as a last resort. They are often very busy taking care of the technical and nontechnical jobs that keep the MOO running, and being interrupted by questions all day long can become frustrating. We emphasize, however, that in case of emergency, a MOO administrator will be more than happy to help you. Don't hesitate to contact one.

2.9 Leaving the MOO

When you are ready to leave the MOO, type '@quit'. This will disconnect you from the character you were using on the MOO. Then all you need do is exit your Telnet program.

3.0 Joining an Education-Related MOO Community

You may already know of a MOO community in which you would like to become a regular participant. Alternatively, you may want to visit a variety of MOOs as a guest to discover the ones that best fit you and your interests. When you do find a community that interests you, the first step in joining is to request a regular character. Most MOOs automate this process and to fulfill your request require only that you have an email address of your own. Other MOOs are designed for use by people from certain real-world communities (such as media researchers or biologists) and may ask for a statement describing your relevant real-world activities. Be sure that you have read any available information about the MOO's theme to know whether regular participants are expected to have any special interests or qualifications.

3.1 Requesting a Character: @Request

One typically requests a regular character on a MOO by connecting as a guest and using the @request command. You may type 'help @request' for information, but the syntax is usually

```
@request <character> for <email address>
```

where <email address> is your own email address and <character> is the name you would like to be called in that MOO community. Note that MOO names cannot include spaces due to certain features of the MOO software.

The @request command tells you if your chosen name is already in use and therefore unavailable. Another way to find out if a name is available prior to @requesting is by typing '@who <name>'. In addition to choosing a name that is not already in use, you may need to determine whether there is a policy about acceptable names. Some large educational MOOs require real name derivatives when possible; a MOO dedicated to the works of a specific author might insist on names used in that author's works, while a more purely social MOO might have no strict naming guidelines. In general, most education-related MOOs will not accept names that the administrators consider profane or otherwise inappropriate for a populace that can include children.

Occasionally the @request command generates an error message and quits before sending your request. The most common cause is that the site you are connecting from does not match the email address given in the @request command, which can happen when you connect through another MOO or a gopher gateway, or when the MOO's nameserver is unreachable. In this case, the best option is to type @wizards to see what administrators are connected, and to page one of them to explain the problem.

Finally, some MOOs require that you give additional information as part of the request process, such as your real name or your research interests. If you are concerned that this information be kept confidential, you may check with one of the MOO's administrators about its policies. MOOs that strive to support collaborative interaction among groups of teachers and researchers, for example, may require that names and addresses be part of the public record so as to facilitate contact between people.

If you are asked to enter additional information and decide you want to cancel the request process, type '@abort' on a single line and press return/enter.

3.2 Connecting to Your Character

When your character request has been granted by a MOO's administrators, they will send a confirmation to the email address you gave as part of @request. It will include a password, of which you should make careful note. The time it takes to get this confirmation varies from MOO to MOO, but do not be surprised if it is a matter of several days (or even longer).

To connect to your regular character, all you need do is follow the procedure you used as a guest. At the welcome screen, type 'connect <character> <password>' and press return/enter. Please note that MOO passwords are case sensitive. You must enter your password exactly as it appeared in your email confirmation.

Welcome to the MOO community!

3.3 Customizing Your Character

The first thing to do once you have successfully connected to your character is change your password from the machine-generated one you received to one of your own choosing. All the usual password-making rules should apply to your choice of MOO password (no family names, pet names, birthdays, or other obvious personal information). Never give anyone else your password or store lists of MOO passwords on computer files that other people can read.

The next thing you will want to do is choose a description and gender for your MOO persona. You do so by using @describe and @gender (as explained in sections 2.4.1 and 2.4.2). The text you use in the @describe command is what anyone will see when they look at you, which explains why one generally uses third person. If you decide to change your description later, you can either @describe again or edit your description using the MOO's line-oriented text editor (see sec. 9.0).

But as a regular character you now have powers for personalization that go well beyond those available to guests. Most MOO objects, including your character, have a variety of default messages associated with them. Many people, for example, choose to customize some of the more common messages that appear when they communicate and move about the MOO. These messages are what you and others will see when you do such things as @go somewhere, @join or page someone, when someone pages you, and in many other circumstances.

The '@messages' command will show you a list of the commands you can use to change the messages displayed by an object (type 'help @messages' for more information). To view your own messages, you can type '@messages me'. Each displayed line consists of the command you would use to set a specific message to its current value. A sample line might look like this:

```
@oarrive me is "%N arrives, waving to all %p friends."
```

The first part of this line, @oarrive me is, is what you type to set the message, but you must also provide an argument. The part enclosed in quotes is the default to which the message is now set, and models the form of any new argument you might wish to use instead. In this example, the oarrive message is what other people in a room see when you enter it: (character) arrives, waving to all (your possessive pronoun) friends. To change this to something else you could type '@oarrive me is "<whatever message you want to use instead>"'.

This example gives us a chance to explain more about what your gender setting is used for. %N and %p are calls for name and pronoun substitution, respectively (type 'help pronouns' for a detailed explanation and a list of available options). The MOO system will parse the messages you set and execute pronoun substitutions when it encounters a substitution symbol. The pronouns it substitutes depend on what your gender is set to. It is possible to do messages without using substitution symbols (for example, by setting the message to "Sky arrives, waving to all her friends"), but using them gives you more flexibility. If you should decide to change your name or gender, for example, there will be no need to edit your messages.

As you will see from the list '@messages me' returns, there are a great many messages you can customize if you want, and little evidence of when each message is used. As you gain more experience, you will begin to recognize the defaults. It may also be useful to know that the convention is to distinguish messages that others will see by prepending an 'o' to a message's name. For instance, @arrive is what you will see when you @go into a room; @oarrive is what will be displayed to other people in that room, @depart is what you will see when you @go from a room, @odepart is what people will see when you leave, and so on.

3.4 Feature Objects

One easy way to extend your character's capabilities is by adding features to it. Essentially, a feature is a type of MOO object that acts as a repository for commands or capabilities that go beyond those basic features already programmed on all MOO players. The creator of a new command (or "verb") can share it with others on the MOO by programming it onto a feature object. Then, in order to use the capabilities enabled by a feature, all people have to do is add that feature to their characters.

Some features automatically come with your character when it is created (type '@features' to see what they are). You may add other features to your character via the @addfeature command (type 'help @addfeature' for more information). See section 4.1.1.2 for an example of an added feature. You can remove a feature from your character with the @rmfeature command.

Most educational MOOs put lists of available features in a central location, such as a generics library or gallery. The location of generics (such as features) would be a good question to ask of a more experienced person on the MOO you are using. You can also see a list of the features that a specific person is using by typing '@features for <character or object number>'. A well-documented feature, like any MOO creation meant for public use, has a customized help message that gives information on what it does and how to use it.

3.5 Finding a Home: @Sethome

On the MOO, your home is where you "wake up" when you connect, and the place to which your character will be moved (no matter where you abandoned it) to "sleep" when you disconnect from the MOO. Traditionally all new users sleep in Limbo, a room with no exits and lots of unembodied characters. When someone connects to one of these characters, it is moved to another room that serves as the starting point for homeless users and guests, and this is where the person wakes up. Now how cozy and inviting does Limbo sound, anyway? No wonder that one of the first things most people want to do is find a new place to call home.

You designate the location you want as your home by using the '@sethome' command while in the room of your choice (type 'help @sethome' for details). Now whenever you type 'home' you will be returned to this place. In addition, when you disconnect from the MOO your character will be moved to your new home, and anyone entering it will see you snoozing away. Note, however, that if a room does not belong to you (e.g., if you did not build it yourself) and has not been configured so that anyone can use it as a home, you will have to get the owner to add you to the room's residents list (via the '@residents' command) before @sethome will work for you.

Some MOOs have hotels, dormitories, or other public residences specifically configured so that anyone can make a home there. Many people also have "cooperative" shared-housing arrangements with friends. Alternatively, you may wish to build your very own place on the MOO. We show you how to begin in section 6.0.

4.0 Interacting in the MOO Environment

MOOs are highly interactive environments, and this interactivity is what makes them so attractive for teaching and learning activities. Although one can use a MOO productively by knowing only the few commands we have described thus far, MOOs offer additional options for interacting with people, places, and things. Knowing about a few of them can make your MOO experience richer.

4.1 More on MOO Communication

One of the truly powerful things about the MOO environment is that it supports mutual awareness (via the @who command, among other things) and a sense of contextualized "copresence" among people remote from one another. But what these things, in turn, support is interpersonal communication—the heart and soul of any MOO. In this section we describe the most commonly used tools for both synchronous and asynchronous MOO communication.

4.1.1 Interacting with Others in the Same Virtual Room

The two most central means of communicating on a MOO have already been mentioned. Recall that one talks to someone in the same room by using the 'say' command, which may be abbreviated as a quotation mark ("). In addition to saying things, one can perform "nonverbal" actions by using the 'emote' command, which may be abbreviated as a colon (:). But there are a variety of other communication tools on a MOO, and some of the most frequently used are described below.

4.1.1.1 Directing Remarks to a Specific Person In multiparty conversations, it is often useful to be able to direct a remark to a certain person. The 'to' command, mow standard on most MOOs, allows this. Its syntax is 'to <character> <message>'. For example:

```
Sky types: to Knek How are you?
Sky sees: Sky [to Knekkebjoern]: How are you?
Others in the same room see: Sky [to Knekkebjoern]: How
are you?
```

In this example, *Knek* is an alias for Knekkebjoern; for more on aliases, see section 6.2.2.

4.1.1.2 Social Features Most MOOs also have certain "social" features that your character can add. As described in section 3.4, features allow you to extend your capabilities on the MOO by expanding the verbs available to you as commands. Social features enable you to perform certain conventional actions (and sometimes rather unconventional ones as well!) simply by typing the name of a predefined action verb (without preceding it by the emote command). As an example:

```
Sky types: smile Knek
Sky sees: You smile at Knekkebjoern.
Others in the same room see: Sky smiles at
  Knekkebjoern.
```

Remember, in order to use the capabilities made available by a feature, that feature must be "on" your character. Use 'help <#feature number>' for information on the verbs that feature provides.

4.1.1.3 @Paste Another very useful communications tool for collaborations on the MOO is the '@paste' command. This command allows you to copy blocks of text from either your MOO client or another program and reprint the text to everyone else in the same room as you. To use @paste in the MOO, you must be able more generally to capture and paste text using the editing features of your client or operating system (e.g., *copy* and *paste*). To paste the text into the MOO you need to capture the text you want, type '@paste', and dump the captured text (e.g., through the paste command on your client's editing menu). You can then terminate the @paste command by typing a single period all alone on a line, and pressing the return key. The pasted text should now be displayed to everyone in the room, including you.

4.1.2 Interacting with People in Different Locations

Almost all of the commands discussed above have parallels that can be used for communicating with people who are not in the same MOO location as you. As already mentioned, a page is the most common means of talking to someone in a different room. The syntax for the page command is 'page <character> <message>'. For example:

```
Sky types: page Knekkebjoern Is it time for the meeting
yet?
Sky sees: Your message has been sent.
Knekkebjoern sees: You sense that Sky is looking for
  you in Sky's Cottage.
She pages, "Is it time for the meeting yet?"
```

In this example, Sky sees the default 'page' confirmation message. Recall that it's possible to customize the messages that someone sees when they page you, or when you page them (as described in section 3.3). The messages you can set that are relevant to paging are named @page_absent, @page_origin, @page_echo, @page_refused, and @page_msg. You can see their default values by typing '@messages me'; use 'help @messages' for more information.

4.1.2.1 Remote Emote On most MOOs, one can send an emote to someone in a different location by typing: +<character> <text of your emote>. Only the person to whom the emote is directed will see it, even if there are others in the same room. For example:

```
Knekkebjoern types: +Sky waves hello from across the
  MOO.
Knekkebjoern sees: Sky has received your emote.
Sky sees: [From Meridian Plaza] Knekkebjoern waves
  hello from across the MOO.
```

4.1.2.2 @Pasteto The remote corollary of the @paste command is '@pasteto'. It is used in exactly the same way as @paste (sec. 4.1.1.3). The difference with @pasteto is that only the person to whom it is directed will see your paste, and they will see it even if they are not in the room with you.

4.1.3 Interacting with People on a Different MOO

A great innovation for users of educational MOOs is the Globewide Network Academy (GNA) InterMOO network, developed and maintained by Gustavo Glusman as part of his work with the GNA. The GNA network links some sixteen MOOs[1] allowing people to see who is on another networked MOO at any given time and communicate with them. To use the GNA interMOO communication channel you must be on one of the GNA-networked MOOs and be able to add the "InterMOO Communication Feature" to your MOO character (the feature will have a different number at each MOO). Take a look at the feature's help message (type 'help <feature#>') for information on how to use it.

The GNA network also provides a meeting place on each MOO called the GNA Forum, in which you can talk to people who are in the corresponding GNA Forum on one of the other networked MOOs. Using the Forum does not require any special features; you need only find the room and enter it. (You might attempt to find its object number by typing '@audit GNA_Architect for "GNA Forum"'. If GNA_Architect owns something by that name, then @audit will show you the object number and you can @go to it.)

Although several other MOO networks have been developed, the GNA network continues to be of primary relevance for education-related uses and users.

4.1.4 Conversing Privately

There are times when one wants to make sure that something is seen only by its intended recipient(s) and not overheard by others in the same room. For one-to-one private communication, the most common method is paging (sec. 2.6.3). If someone is in the same room as you, you may also communicate privately by using the 'whisper' command: 'whisper "<message>" to <character>'. (Recall that '@pasteto' is also a private communication; see section 4.1.2.2.)

The options are more limited for multiparty conversation. Some MOOs provide the equivalent of "chat lines," which resemble CB channels or, more recently, Internet Relay Chat (IRC) channels. If the MOO you are using offers such an option (usually enabled by adding a channel feature), it should be possible to create a restricted, locked channel (see sec. 7.4) that only the people you designate can use. You may need to ask around to find out whether this feature is available on the MOO you are using.

Of course, another option for holding a private conversation is to ensure that only the people you want to speak with are in the room with you. You should be aware that planting "bugs" and finding other ways to listen in on conversations are possible on a MOO, although doing so is a very serious offense. If it is important to you that the group's conversation not be overheard, then it is worth using a few of the MOO's security measures if you choose the "same room" option.

Typing '@sweep' will tell you who is able to listen to normal conversation in the room you are in. Make sure that only the names you expect are on the list that @sweep gives you. After sweeping the room, it is prudent to "lock" it to prevent others from entering. The commands for enabling and disabling room security are implemented differently on different MOOs, so consult the online help to find out more about your options for locking a room. In general, you can only lock a room that you own. Remember to unlock the room when you no longer want privacy, because usually not even invited guests can enter it until you do.

Please see section 7.0 for more information on these and other security measures.

4.2 Asynchronous Communication

The commands discussed thus far pertain to "live" interaction on a MOO. MOOs also support asynchronous communication among people who are not connected to the MOO at the same time, or by someone who wants the recipient(s) to be able to read a message later.

4.2.1 MOOmail

A MOO supports its own internal electronic mail system, typically referred to as MOOmail to distinguish it from regular email. The MOOmail system allows you to send and receive messages to and from other people

TABLE 2. Basic MOOMail Commands (use 'help <command>' for more information)

Command	Description
For reading mail	
@mail	Lists headings for the contents of your mail folder; can also be used to search a collection of mail.
@read	Shows a mail message (this command can take a variety of arguments, such as @read <msg #>).
@next	Shows the next message in your mail folder. Also check out @nn.
@prev	Shows the previous message.
For sending mail	
@send	Initiates message composition. Check the use of the @enter command within @send. Use 'quit' or 'abort' to exit the mailer before sending a message.
@answer	Replies to a message in your folder.
@forward	Resends a received message to someone else.

on the MOO, whether or not they are connected when you are. It allows you to save messages, view and manage your mail collection, and, in some cases, forward MOOmail to your regular email address. It also allows you to read and contribute to any public mail collections that are available on the MOO (such as mailing lists).

The basic MOOmail commands are shown in table 2. Note that most of these can take arguments on the command line (e.g., '@read <number> <on *collection name>', instead of just '@read'). In-MOO help is available on all of the commands that we mention here, and we encourage you to turn there for details about the MOOmail system and how to use it.

MOOmail is a great advantage, but mail collections can take up a lot of memory on a MOO (see sec. 5.1 on quotas), and unless you delete a message it stays in your MOO collection. For this reason, you may want to be diligent about deleting old or unimportant MOOmail. Check the online help for @rmmail/@unrmmail to learn about deleting MOOmail messages, and for @netforward if you would like to send MOOmail on to your regular email address before you delete it. It is also possible to customize the behavior of the MOOmail system to suit your preferences, although the way in which one does so is rather complex. Type 'help @mail-options' for more information.

TABLE 3. Useful Commands for Viewing in-MOO Mailing Lists

Command	Description
@subscribe	Shows the names of all the mailing lists that exist on a MOO.
@subscribe <mailing-list name>	Subscribes you to the named list. You will be notified when you connect if there are new messages on that list.
@subscribe <mailing-list name> with notification	Same as preceding command, except you will be notified immediately when someone sends a new message to the list.
@subscribed	Show all the lists that you have subscribed to.
@unsubscribe <mailing-list name>	Ignore the named mailing list from now on.
@rn	Shows the collections that have new messages on them.
@skip <mailing-list name>	Ignore any unread messages in the named list.
@send <*listname>	Send mail to the named list.

4.2.2 In-MOO Mailing Lists

Most MOOs also provide for a variety of public (and some private) mailing lists to which you can subscribe or read as you desire. Type '@subscribe' to see what mailing lists are available. Table 3 shows a few commands that are useful for viewing these lists, but be sure to check the online help for more detailed information on how to use each command more flexibly. (Mailing lists are commonly referred to as *mail collections* in the online documentation, just like an individual's undeleted MOOmail. Note also that mailing-list names begin with an asterisk—e.g., *general.)

4.3 Interacting with the MOO Environment

There is a lot to be seen and explored in a good MOO. At the simplest level, one can interact with things in the MOO environment simply by looking at them: 'look <object name>'. When you look at something, you see the description its creator has written to represent it within the virtual reality of the MOO. Well-designed places and things may incorporate ever-deepening layers of detail for you to look at. If a room's description mentions a painting, you may be able to type 'look painting' and see the painting's description; if, the painting's description mentions the figure of a small boy, you may be able to type 'look boy' and see a description of him.

But one can do far more than look around in a MOO. Many objects in the MOO world have additional things you can do built into them, and these actions will produce interesting effects and messages of the creator's design. Although one could use trial and error to discover the possibilities for interacting with something, another way is to look beneath the surface using the '@examine' command. As with most other commands that begin with the @ symbol, @examine lets you step outside the virtual reality, in this case to get more information on the level of mechanics.

If the object (or person) is in the same room as you, you can just refer to it by name: '@examine <character or object name>' (e.g., '@examine bookshelf') will show you the verbs (or commands) you can apply in relation to that person, place, or thing, along with a great deal of other information about it. If the object is in a different location, you need its object number (see section 5.5 of this chapter for some tricks for finding object numbers).

Let's use a fortune teller we once created for a Halloween party at Meridian as an example. Typing '@examine fortune teller' shows the following:

```
Fortune Teller (#10153) is owned by Sky (#79).
Aliases: Fortune Teller and teller
An ancient crone who knows many secrets, she sits here
  behind the curtained opening of her tent, waiting for
  those who dare to <consult teller> about their
  destinies.
Key: (None.)
Obvious Verbs:
  consult fortune teller
  @addfort fortune teller
  @listfort fortune teller
  @rmfort <anything> from fortune teller
  g*et/t*ake fortune teller
  d*rop/th*row fortune teller
  gi*ve/ha*nd fortune teller to <anything>
```

The list of obvious verbs shows commands you can use to interact with the fortune teller. Typing 'consult fortune teller', for example, would produce an amusing sequence of messages in which she gives you a uniquely MOO reading.

Interacting with objects in the MOO environment is great fun. There are many, many different kinds of objects—some created with serious intent, some for educational purposes, and some just for amusement. Creating objects that have entirely novel commands (aka verbs) associated with them requires that you learn a bit about MOO programming, which can be great fun in itself (see sec. 8.0). But even if you are not yet a programmer, you can still make your own fun and useful objects on the MOO and customize their behavior in a variety of ways.

5.0 Creating Objects

One can quite literally change the world in a MOO by creating and customizing objects that will continue to exist on the MOO, even when you are not connected. In this section, we tell you how to create your own objects, beginning with some important background on resource management and the nature of MOO objects. Building new places is a special form of object creation central to many people's MOO experiences, and we will discuss building as well as creating. We will also describe some of the commands that help you keep track of your objects and get rid of ones you no longer want.

5.1 Quota

Like any other computer program, a MOO requires a certain amount of the server machine's memory and processing resources to operate. Each and every object in a MOO occupies a small chunk of memory. If there were no system for managing a MOO's growth, then that MOO could quickly outstrip the resources of all but the most computationally fortunate host systems. It would run ever more slowly and eventually crash.

The quota system is the MOO's solution for managing limited resources that must be shared among a large community of people. Your new character will typically have been allocated a certain quota, and that amount determines how much you can create and build on the MOO. The '@quota' command shows the quota you have available. For each object you create your remaining quota is lowered, and for each object you recycle it is raised. The size of the standard quota allocation is determined by each MOO's administrators, and they also make their own policies about increasing a character's quota.

MOOs originally used an object-based quota system, meaning that each user was allocated a certain number of objects. A quota of ten meant that you could add ten new objects to the MOO. The problem is that not all objects are of equal size. Some incorporate elaborate verbs and properties, each of which requires memory in the MOO database. Therefore, one person might own ten objects that together take up two thousand bytes of memory, while another might own just one object that takes up four thousand bytes.

To manage resource use more fairly, many MOOs now use a byte-based quota system. In such a system, people are allocated a certain number of bytes of memory they can use. This system is a better and more flexible reflection of true memory usage, and is implemented in the core database that supports this book.

The bottom line is that to add something new to a MOO, you must have some quota available to you. Type '@quota' to check. If you need more memory for a new project, page a MOO helper or another knowledgeable person and ask how to apply for a quota increase.

5.2 The Object Orientation of MOO

A MOO is an *object-oriented* environment. At the simplest level, this means that most everything on the MOO (including your character) is an object and has a unique object number assigned to it. At another level, however, object-oriented describes the programming environment from which a MOO is built and by which MOO users can extend it. Rather than starting from scratch each time we want to create something in an object-oriented environment, we say what existing "generic" object we want the new object to resemble. The new "child" object automatically inherits the capabilities and behaviors of the generic "parent" object, and all that remains is to customize the new object as we desire.

Every object on the MOO is created in this way, so every object has a parent object from which it inherits family traits. That is, each object begins with the characteristics of its parent, and of its parent's parent, and so on, going all the way back up to the granddaddy from which all else descends: object #1 (the root class object). The root class object provides all the objects in a MOO with a common set of behaviors, such as the ability to have a description and a name. Further down the hierarchy from object #1 we find the generic thing (object #5). Now, in addition to the capacity for a name and a description, this object has behaviors we expect from "real" things: we can drop it, move it, give it away, and pick it up.

As one continues to descend through the object hierarchy, the generics come with more and more specialized behaviors. There are standard classes, such as generic containers, notes, and so forth, as well as more specialized generics that have been contributed to a particular MOO by its members. The happy consequence of this, as we have said, is that when you want to create an object that resembles, say, a television, all you need do is find a generic object that has good characteristics for serving as its parent. Many MOOs put lists of their generics in a central location, such as a generics library, where people can browse through the collection to see what is available.

The @parents command is useful for displaying the genealogy of an object and getting an idea of the different behaviors it inherits from its predecessors. To illustrate, let's consider a generic TV, with object number, let us say, 117. The command '@parents #117' returns the name and object number of the generic TV itself (#117), the generic thing object (#5), and the root class object (#1). If you decide to create a child of the generic TV, you are able to give it a name and a description thanks to its descent from the root class object (#1). You are able to drop it and give it away to other users because it inherits those capacities from the generic thing (#5). And because its parent is the generic TV (#117), you are able to do "TV-like" things to it, including turning it on and off and choosing channels.

Welcome to the object-oriented world of MOOs!

5.3 Making a New Object: @Create

To make a brand new MOO object, use the @create command:

```
@create <name or number of parent object> named "<new
  object name>"
```

Let's return to the TV example we used a moment ago. To create your own TV named MyTV, type '@create #117 named "MyTV"'. The MOO responds something like this:

```
You now have MyTV with object number #154 and parent
  Generic TV (#117).
```

With this act of creation, you have brought a new object into existence on the MOO. You can describe it, customize it, drop it, and so forth because you are its owner, as you will see if you type '@audit.' The '@audit' command usefully lists all the objects you own ('@audit <character>' will list someone else's things); type 'help @audit' for more information on this command. The object number of MyTV is made up for this example, of course, because the number assigned to a new object depends among other things on how many things already exist in a particular MOO.

Typing 'inventory' shows the things your character is carrying around, and you might be surprised to discover that you are now carrying your new TV (not a bit heavy, is it?). A newly created thing is automatically placed in your pocket, as it were. If you drop it or hand it to someone else (for example), it no longer appears "on your person." The things you are carrying are shown to other people when they look at you.

5.4 Customizing Objects

All objects can be customized. At the simplest level, they can be named and described, but many also allow you to specify messages to display at certain times, or to customize their reactions to the environment. There are many parameters that the creator of a generic object can make available for customization, so your choices depend heavily on what generic parent you use when creating your object. In general, you can use '@examine' to see the syntax for commands specified on your objects. You can also check '@messages <object name>' to see if there are any special messages to be set, and use 'help <object name>' to see if the creator has provided some additional guidelines. If you have programming privileges on a MOO, you have additional options available for seeing the inner workings of an object and for endowing it with entirely new characteristics (see sec. 8.0).

5.5 Tracking, Moving, and Recycling Objects

As already noted, the '@audit' command is handy for keeping track of the objects you own, where they are located, and how much quota they use. If you have misplaced something, @audit will show you where it is, and you should then be able to use the @move command to get it back. For example, '@move <object number> to me' will put an object into the inventory of things you are holding; '@move <object number> to here' will move it to your current location.

Once you no longer want or need one of your objects, it is time to recycle it: '@recycle <object name or number>'. The @recycle command irrevocably destroys the object (and any changes or customization you have done to it), and put its object number back into the pool for assignment to new creations. To minimize the chances of recycling something by accident, the '@recycle' command asks you to confirm your decision before it proceeds. If you respond with a 'yes', the object is then sent to a recycling bin where it is stripped of its former personality and made ready for reuse. If you accidentally recycle something of extreme importance, it might be possible for one of the administrators to get it back for you by activating a backup of the MOO database and retrieving it from that. This entails a fair amount of work, however, so be sure it is not easier just to recreate the object from scratch. In general, it is a good idea to keep backup copies of your important creations somewhere outside of the MOO. You can then recreate your MOO things by using your backup in case they are ever lost in a MOO crash or recycled by accident. How to do offline backups is another good question to ask of a more experienced person on the MOO.

6.0 Building New Places

Building a new place is effectively just another form of creation, but building is done with the '@dig' command instead of '@create'. The @dig command makes building simpler by assuming that the parent is a generic room (and thus have the characteristic behavior of a room), and by providing you with the option to make exits for the new room at the time you create it. Although digging a new room can be as simple as typing '@dig <new-room-name>', this will merely create a disconnected room with no conventional way in or out. Therefore, before we explain precisely how to use @dig in more detail, let's look more closely at how exits and entrances contribute to a well-built place.

6.1 Exits and Entrances

Rooms on a MOO are another type of object, in this case with a parent that traces back to object #3, the generic room. But rooms typically need to have exits that connect them to other rooms on the MOO, and these exits, too, are objects with their own parentage (tracing back to object #7, the generic exit) and object numbers. Each exit is a one way passage; that is, digging an exit creates an object that goes only *from* the room of origin to a destination. It is not usually good form to create a room with a way in but no way out (or vice versa), so you need two exits to make the usual two-way connection between rooms: an exit from room A to room B, and an exit from room B to room A. These exits could in principle be named "Charles" and "Diana." In practice, however, one should name them in a way that makes sense for people trying to navigate through the MOO. Direction names (and/or abbreviations) are the traditional choice in a MOO, as in 'north' 'n', 'northeast', 'ne', and so on. If such compass directions are used for names, however, then it is also considered good practice to use complementary names for the two exits. In other words, if one uses 'east' as the exit from room A to room B, then one should use 'west' as the exit from room B to room A. This is because people draw on their experience with real world spaces to navigate in the MOO. They expect to be able to go back the way they came, simply reversing their course.

Conventional exits are brought into being via the @dig command, as explained below. They may be created at the time a room is first dug (via arguments to @dig) or may be added later. Keep in mind that only the owner of a room (or an administrator) can attach an exit to it. Therefore, if you want to make a return exit from someone else's room to yours, you most likely need to get the other owner to do it (see sec. 6.2.3).

6.2 @Dig

The @dig command is the de facto way of creating new rooms with or without exits and is capable of doing several different things depending on what arguments you use. The main forms of @dig are shown below, and the remainder of this section is devoted to explaining each in more detail.

```
@dig <new-room-name>
@dig <new-room-name>, "<additional-alias>",
  <additional-alias>
@dig <exit-specifications> to <new-room-name>
@dig <exit-specifications> to <old-room#>
```

6.2.1 The Basic Building Tool: '@dig <new-room-name>'

The most basic form of the command, '@dig <new room name>', creates a room with the given name. The room will be floating in space, disconnected from everything else on the MOO. The @dig command will show you the room's object number, though, so you can still get there with '@go <room #>' or '@move me to <room #>'; if you forget the room's object number, it will show up in your @audit list (though not in your inventory because new rooms are not automatically placed "on" your character). If you wanted, you could even use @move to place the room somewhere else. As a minor word of warning, however, note that if you decide to carry the room around, the MOO will not allow you to enter it because it would then be logically impossible to figure out whether you are inside the room or it is inside of you.

6.2.2 Specifying Additional Aliases for a Room

Rooms and other things on a MOO cannot have more than one name at a time, but they can have aliases that can be used in commands instead of the name. To provide aliases for your room at the time you first @dig it, use '@dig <new-room-name>, <additional-alias>, . . . ,<additional-alias>' (the dots represent additional aliases). For example, typing '@dig Dormitory Building, dorm, dorms, building' creates a single room named Dormitory Building, with three aliases. Adding or removing aliases after a room has been dug can be done with the @addalias and @rmalias commands, respectively (check the in-MOO help for more detailed information).

6.2.3 Creating Exits When Digging a New Room

Another variant of the @dig command allows you to create exits that link the room you are currently in to another room. There are subvariants of this command that differ according to how many of the command's optional arguments you use.

One version lets you make an exit from the room in which your character is currently located to a new room B at the same time you are bringing room B into existence. The syntax to do this is:

```
@dig <exit-specifications> to <room# or room-
  name(s)>
```

By <exit-specifications> we mean the name (and optional aliases) you want to give to a new exit that will go from the room you are in to the new room. You also have the option at the same time to create a second exit from the new room back to the room you're in by separating name(s) in <exit-specifications> with the symbol | . Sound complicated? Not really. Let's go through some concrete examples.

Example 1:

```
@dig north to Blue Parrot Bar
```

Creates a new room named Blue Parrot Bar.

Creates an exit named north from the room you are in at the moment (aka, the current room) to the newly created Blue Parrot Bar.

Example 2:

```
@dig north, n to Blue Parrot Bar, parrot, bar
```

Creates a new room named Blue Parrot Bar, which has the aliases of parrot and bar.

Creates an exit from the current room to the newly created Blue Parrot Bar. The exit is named north and has the alias n.

Example 3:

```
@dig north, n | south, s to Blue Parrot Bar, parrot, bar
```

Does all that example 2 does,

Also creates an exit from the newly created Blue Parrot Bar back to the current room. The return exit is named south and has the alias s.

So in this form of the @dig command, all names before the symbol | pertain to the exit leading from your current location to the other room, while names after the symbol | belong to the exit that leads from the other room back to your current location. (Phew! That's a mouthful.)

It may seem puzzling at first that even though @dig happily creates an exit (or exits) for you in the manner we have described, you would not yet have a working exit unless you own the room it is exiting from. If you @dig from a room you do not own, your new exit will be brought into existence but would not yet work. You will have to get the owner of the room to use '@add-exit <your exit #>' (or '@add-entrance') before your exit will register as connecting the two rooms and function normally. If you forget to make a note of the exit number in question when you @dig it, you should be able to @audit yourself to find it when you need to tell the other owner what object number to use in @add-exit.

6.2.4 Connecting One Existing Room to Another Existing Room

The @dig command is also used to create a new passageway (either one-way or two-way) between two already-existing rooms on a MOO. The syntax is just as in section 6.2.3, except you substitute the object number of an existing room:

```
@dig <exit-specifications> to <old-room #>
```

For example, typing '@dig north, n to #115' creates an exit named north (with the alias n) from your current location to an already existing room with object number #115.

7.0 Basic MOO Security Features

You may intend the things you create and build on a MOO to be freely available for others to interact with as they wish. But, there may be times when you want something of yours to stay where you put it and not be carried off by an enthusiastic explorer who knows how to pick things up but not drop them. That's one good reason to learn about security features on a MOO. At times you may want to prevent people from dropping into a room while you are in the middle of a class session or meeting. And all MOO users should know about the basic "self-defense" mechanisms that exist for dealing with persistent and unwanted interactions on a MOO. Here we discuss some of the options available to you.

7.1 @Gag

It is possible that there will be times when you would rather not hear (or, literally, see on your screen) any text that comes from a particular MOO user or MOO object. Perhaps the object is responsible for a flood of irrelevant text across your screen, commonly called "spam" (chatty MOO robots can be notoriously "spammy," for example). Silencing the user or object with @gag will stop any textual output associated with it from showing up on your screen. You will be rendered unaware of its presence (or absence) in a room or on the MOO. The syntax for @gag is simply: '@gag <object #>'.

A list of things that you have gagged can be used by typing '@listgag'. You can keep things gagged for as long as you wish. To remove someone or something from your gaglist, use the @ungag command.

The conventions under which gagging is used (and therefore, the social implications of gagging someone or being gagged) differ from MOO to MOO. Typically, gagging is more common on large, purely social MOOs than on smaller, more educationally oriented ones. In our combined years of MOOing, for example, neither of us has encountered a situation in which we have needed to @gag someone (though each of us has found it useful to gag noisy MOO objects from time to time). On the other hand, if you do find yourself in a problematic situation, then @gag is a good solution simply to make the problem invisible.

7.2 @Security

Everyone needs his or her "personal space" sometimes, even in a virtual world. Sometimes it's desirable to restrict access to a room, allowing some users to enter freely while keeping out the rest. One option is to use the @lock command, as we describe below for objects in general. In the case of rooms, however, @lock is an awkward way to restrict access by people because it prevents objects from being dropped in the room as well. For this reason, other room security mechanisms have been developed on many MOOs.

Room owners have a set of commands available with which to control access to rooms. By default, sets of people are always able to enter a room: the room's owner, anyone he or she has added to the room's residents list (via @residents), and the MOO's administrators. The owner can prevent access by anyone else by typing '@security on' while in the room. This ensures that anyone trying to enter is checked against an accept list. You can add someone else to the room's accept list via the @accept command; the @reject command will take someone off the list. Typing '@security off' is equivalent to allowing anyone to enter freely, which is the default. You can check who is in your room's access list at any time by typing '@accept'.

7.3 @Eject

The above measures ensure that you know who is allowed into a room but are of no use if an unwanted visitor enters the room before you can restrict access with '@security on'. If someone is unable or unwilling to leave your room, you may wish to type '@eject <character>'. The @eject command immediately expels anyone but a MOO administrator from a room that you own.

7.4 @Lock

Every now and then it is nice to be able to lock an object in place so that it cannot be moved by someone else unless a certain criterion is met. Any object on a MOO can be locked via @lock. One can use it to restrict who can use an exit and when. One can @lock a note object in such a way that only the right person, in the right situation, can read the encrypted message. These are just some of the uses of the @lock command, which is a multipurpose restriction mechanism based on keys of logical expressions and object numbers (or, sometimes, names). In-MOO help is available by typing 'help locking' and 'help keys'. The general syntax for @lock is

```
@lock <object> with <key expression>
```

For the sake of demonstration, let's return to our sample object MyTV (#115). Imagine that you want to place it in a specific room and, since it is a sizable object, you do not want someone to be able to pick it up unless he or she has an appropriate tool—perhaps that wheelbarrow you so conveniently left in the corner. To achieve this little enhancement to the virtual-reality experience, you can lock your TV like this:

```
@lock MyTV with wheelbarrow
```

Now your TV is securely in place until someone picks it up after first having picked up the wheelbarrow. It's that simple.

Perhaps you decide on second thought, however, that you do not want to have to go through the trouble of using the wheelbarrow yourself. It *is* your TV, right? In this case, all you need do is change the @lock key appropriately:

```
@lock MyTV with me || wheelbarrow
```

The "||" is one of several logical operators that can be used with @lock, in this case signifying that either you *or* the wheelbarrow is necessary for picking the TV up. (The set of allowed logical operators in a lock key are '&&', '||', and '!', or alternatively, 'and', 'or', and 'not', respectively. For a detailed description please read 'help keys'.)

Finally, to illustrate some of the flexibility you have with @lock, let's imagine that you reconsider a final time and decide you really do not want to let Phil, your pesky neighbor, pick the TV up at all, wheelbarrow or no wheelbarrow. This is easily accomplished with:

```
@lock MyTV with !phil && (me || wheelbarrow)
```

Now Phil cannot take your TV anywhere, and other people can do so only if they have the wheelbarrow first. You, on the other hand, can pick it up any time you wish. Isn't virtual life great?

8.0 Programming Enhancements to the MOO Experience

Everything on a MOO is built from a programming language called LambdaMOO. When you use a command, you are using a "verb" that has been programmed onto an object by someone else—by the developers of the MOO core, or by another member of your MOO community. We hope there will come a day when you will be so interested in the MOO's potential that you will want to add an entirely new verb to an object or perhaps create an entirely new sort of MOO tool. You will do so by learning the MOO language and by becoming a programmer on the MOO. Ask a more experienced person or a MOO administrator about the circumstances under which the MOO grants programmer capabilities to people. Many education-related MOOs view programming as a great opportunity for teaching and learning and will grant programmer status simply for the asking.

Many people learn MOO as their first programming language, and it has been used successfully by educators at high-school and university levels to teach object-oriented programming. It is a relatively simple language to learn and a very rewarding addition to the MOO experience. A detailed explanation of the MOO programming language is beyond the scope of this chapter, but an excellent manual, *The LambdaMOO Programmer's Manual,* has been written by Pavel Curtis and can be retrieved from the High Wired enCore web site.

All common MOO verbs are declared on objects using the MOO programming language. The data to which these commands refer are held on objects in named *slots*, called properties, as arbitrary MOO values. In the remainder of this section, we will introduce some of the most commonly used commands for handling verbs and properties, as starting points for exploration on your own. For more detailed explanations of each command you can type 'help <command>', although the online help for programming-related commands does presume a certain level of familiarity with basic programming constructs. If you have no prior programming experience, then you will want to spend some time with one or more of the many programming tutorials that have been written and made available on various MOOs or on one of the many MOO-related web sites now in existence. A gentle introduction to MOO programming has been included in the enCore MOO database.

Remember, your character will need to have programmer status (aka, a *programmer bit*) to program MOO verbs and to use the commands we describe in this section.

8.1 @Display

```
Syntax: @display <object>. [property]
                          , [inherited_property]
                          : [verb]
                          ; [inherited_verb]
```

This command displays information about the verbs and properties defined on an object itself and on its predecessors. In this and the remaining examples, the bracketed arguments are optional.

Examining verbs that others have written is a great way to learn. Typing '@display myTV.name' will show information about the name property on your TV object, while '@display myTV.' will display information on all the properties declared on your TV object (or if none have been declared directly on it, the properties declared on the parent object will be shown). To view all the verbs declared on the object plus all the ones inherited from its predecessors, type '@display myTV;'. Examples:

```
Sky types: @display myTV.name
Sky sees:  ,name      Built in property      "myTv"

Sky types: @display myTV.
Sky sees:  myTv (#154) [ ]
Child of Generic TV (#117).
```

```
Location Sky (#100).
on              Sky (#100)    rwc    1
channel         Sky (#100)    rwc    11
index           Sky (#100)    rwc    1
pause           Sky (#100)    rwc    0
tapenum         Sky (#100)    rwc    #1
designers       Sky (#100)    r c    "cdr@mediaMOO"
library         Sky (#100)    r c    #118
studio          Sky (#100)    r c    #994

Sky types: @display myTV:turnon
Sky sees: #117:turnon  Wizard(#2) r d   this none
  none
```

8.2 @Verb

```
Syntax:
@verb <object>:<verb-name(s)>
@verb <object>:<verb-name(s)> <dobj> [<prep>
  [<iobj>]]
@verb <object>:<verb-name(s)> <dobj> <prep>
  <iobj> <permissions>
@verb <object>:<verb-name(s)> <dobj> <prep>
  <iobj> <permissions> <owner>
```

This command adds a new verb, called <verb-name(s)>, to <object>. If you want the verb to have more than one name, separate each name by a space and enclose the whole set in double quotes (as in the second example below). The remaining parameters for @verb are explained in detail in *The LambdaMOO Programmer's Manual.*

Examples:

```
Sky types: @verb myTV:adjust this none none
Sky sees:  Verb added (0).

Sky types: @verb myTV:"adjust fiddle" this none none
Sky sees:  Verb added (1).
```

8.3 @Property

```
Syntax:
```

```
@property <object>.<prop-name>
@property <object>.<prop-name> <initial-value>
@property <object>.<prop-name> <initial-value>
  <permissions>
@property <object>.<prop-name> <initial-value>
  <permissions> <owner>
```

This command adds a new property, called <prop-name>, to <object>. If <initial-value> is not supplied, it will be set to '0'.

Examples:

```
Knek types: @property myTV.bumps
Knek sees:  Property added with value 0.

Knek types: @property myTV.ouch_msg "OUCH! That hurt!"
Knek sees:  Property added with value "OUCH! That
            hurt!".
```

8.4 @Program

```
Syntax :
@program <object>:<verb-name>
@program <object>:<verb-name> <dobj>
  <preposition> <iobj>
```

This command allows you to enter code for a verb and compile it. (To change a verb program and compile again, you can use the editor: '@edit <object>:<verbname>'.) It will replace any existing code of the verb <verb-name> declared on <object> (if there was any) unless you @abort in the middle.

When you use this command, everything you type is assumed to be lines of MOO code (so that any MOO commands that you are specifying would not be executed, for example). Typing '@abort' by itself in a line (and pressing the carriage return) will immediately exit this line mode and quit @program without attempting to compile. Typing a period in a line by itself will also exit line mode, but then @program will attempt to compile the lines of code you just entered. If your code has any errors, you will get an error message, pointing to what the MOO thinks the problem is. If there are no errors, you will get the much-beloved message that your verb has compiled successfully. Example:

```
Knek types: @program myTV:adjust
Knek sees:  Now programming myTV:adjust(0).
            [Type lines of input; use '.' to end or
            '@abort' to abort the command.]
Knek types: player:tell(this.ouch_msg);
            this.bumps = this.bumps + 1;
            player:tell(this.name + " now has " +
            tostr(this.bumps) + " bumps.");

Knek sees:  0 errors.
            Verb programmed.
```

Please read *The LambdaMOO Programmer's Manual* for a detailed explanation of the MOO programming language and what the above lines do.

Incidentally, if you have tried any of these examples, then you now have a new verb on your TV object. If you added the two properties .ouch_msg and .bumps, then you could now use your verb to interact with your TV in a new way:

```
Sky types: adjust myTV
Sky sees:  OUCH! That hurt!
           myTV now has 1 bumps.
Sky types: adjust #154
Sky sees:  OUCH! That hurt!
           myTV now has 2 bumps.
```

Remember, when you want to make changes to a verb it is usually much simpler to use the verb editor. See section 9.0 for more information.

8.5 @List

```
Syntax: @list <object>:<verb>
        @list <object>:<verb> [with|without
        parentheses|numbers] [all]
        @list <object>:<verb> <dobj> <prep>
        <iobj>
        @list <object>:<verb> <start>..<end>
```

This command prints the code of <verb>, declared on <object>. <start> and <end> indicate line numbers. Each line of code is prepended with its line number, unless the 'without numbers' parameter has been provided. So this command is useful for taking a look at the code of (readable) verb.

```
Knek types: @list myTV:adjust
Knek sees:  #154: "adjust fiddle"    this none none
                1: player:tell(this.ouch_msg);
                2: this.bumps = this.bumps + 1;
                3: player:tell(this.name + " now has " +
tostr(this.bumps) + " bumps.");
```

9.0 The In-MOO Editors

Imagine that you have just finished entering the perfect description for your new room, only to discover (after you have pressed the return/enter key, of course) that a few typographical errors snuck in. You could redo the whole @describe command, but a better answer in this and a variety of related situations is a text editor.

Because the MOO is a line-based environment, its editors are, too. If you are old enough to remember UNIX line editors (e.g., vi and ed), then you will know what to expect from the in-MOO editors. If you are not, then suffice it to say that these editors are not necessarily going to bring the words *efficient* and *user friendly* to mind. Nonetheless, even though they may take a little getting used to, the MOO's editors effectively allow you to update things and fix mistakes without having to start all over.

9.1 Starting an Editing Session

There are three primary MOO editors, one to change MOO programs (the verb editor, @edit), one for working with text objects (the note editor, @notedit), and one specialized for handling mail messages. These editors are all descendants of the generic editor object and function in much the same way.

In the remainder of this section, we will concentrate on the verb and note editors. On most MOOs, you can invoke either by using the command '@edit <whatever you want to edit>'. On these MOOs, you will then be placed in the editor appropriate for the type of object you specify. Just in case, however, the original commands for invoking these different editors are as follows:

```
@edit     <object>:<verb>     invokes the verb editor
  (edits verb code)
```

```
@notedit <note_object>          invokes the note editor
  (edits note text)
@notedit <object>.<prop>        invokes the note editor
  (edits text property)
```

When you successfully start an editing session by using these commands, your character is transported to a special (and solitary) editing room. If you were in a room with other people when you invoked the editor, you will no longer be able to hear or be heard by them (though you may still page). When you leave the editor, you will be returned from where you came.

9.2 Editor Commands

Within either editor, typing 'look' will list the editing commands you have available. Typing 'help <command>' for any listed command should display a short explanation and the correct command syntax. There are many commands available in the editor, and we will only touch on a few of them here.

The main thing to remember is that you must deal with the text you are editing on a line-by-line basis. Anything you tell the editor to do applies to the current line only—that is, "the current insertion point." When you start an editing session, the insertion point is after the last line of existing text. To change it, you must move it up (or, later, down) by using the appropriate command (e.g., prev). The basic way you add a new line is by "saying" it; that is, you precede each new line you want to add by a quotation mark. That line will be inserted at the current insertion point (of course). If you then add another line, it will be placed just beneath the previous one, and the pointer moved down accordingly. Similarly, if you want to change something on a particular line, you must first give the right commands to move the pointer to that line. You can see a group of lines by typing 'list' and can then move the pointer to the one you want.

Let's look at an example of using the editor to put text on a sign one might want to have in the Blue Parrot Bar. First, we will create an object whose parent is a generic note, so that we can "write" on it: '@create $note named "Sign"'. We can '@describe the sign as "A hand-lettered sign, hanging on the wall of the bar. 'read sign' to see what it says."' If you look at the sign, this description is what you would see. But if you tried to 'read sign', there would be nothing written on it. We need to put something there, and we will do that by invoking the note editor with '@edit sign'. This causes us to be moved to the note editor room, where next thing we see is:

```
    Text Editor
Do a 'look' to get the list of commands, or 'help' for
  assistance.
Now editing "sign" (#18743).
```

Because this object does not yet have text on it, typing 'list' will simply inform us of that. If there had been something written on this note already, then typing 'list' would have shown us some or all of the existing lines of text, numbered.

One way to enter text in this room is with the say command, terminating every line with a carriage return. For example,

```
You type: say Welcome to the Blue Parrot Bar, where we
  serve
You see:        Line 1 added.
You type: say every kind of exotic juice drink you can
  imagine.
You see:        Line 2 added.
You type: say Literally.
You see:        Line 3 added.
```

Typing 'list' now shows you the three lines of text, and the insertion point is below line 3. One can now type 'save' to write these changes onto the sign. If there is nothing else to add or change now, typing 'quit' will return you to the room you were in before beginning this editing session. You could now type 'read sign' and see the following:

```
Welcome to The Blue Parrot Bar, where we serve
every kind of exotic juice drink you can imagine.
Literally.

(You finish reading.)
```

There is a very useful alternative method for entering text in the editor, the command 'enter'. With it, you no longer have to precede every line with a 'say'. Instead you can type or paste the text directly into your note object. When you are done entering your description, you can return to normal editor mode by typing a line whose sole text is '.', to keep your new changes for now, or '@abort', to discard anything you might have written. If you choose this option, setting the description might look like this:

```
@edit sign
Now editing "sign" (#18743).
enter
[Type lines of input; use '.' to end or '@abort' to abort
  the command.]
Welcome to the Blue Parrot bar, where we serve nearly
  every kind of juice drink you can imagine. Literally.
  All you have to do is imagine what you want to drink and
  pretend to drink it!

Enjoy!      --The Management

Lines 1-3 added.
save
quit
```

Because you did not press the return/enter key until after you typed 'it!', the editor treats everything preceding that as one single line. The blank space, which you get by pressing return/enter again, is also counted as a line of text.

But let's imagine that you are not quite satisfied and decide you would like to change a few things on the sign before hanging it in the bar. Once again, you use the editor by typing '@edit sign'. The insertion point starts out on the last line of text. Let's say you want to change "--The Management" to something else. For this you need to use the editor's syntax for substitution:

```
You type: s/The Management/The Proprietor/
You see:          ___3_ Enjoy!   --The Proprietor
You type: save
You type: quit
```

And now your sign will read:

```
Welcome to the Blue Parrot bar, where we serve nearly
  every kind of juice drink you can imagine. Literally.
  Just imagine what you want to drink and pretend to
  drink it!

Enjoy!      --The Proprietor
```

You can get a complete explanation of how to use substitution by typing 'help subst' while in the editor. For a list of other commands, type 'help commands' then 'help <command name>' for detailed information. The editors have commands to search, substitute, copy, delete, append and more. The verb editor has additional commands, like 'compile.'

The MOO editors provide all users with basic editing capabilities, and for those with no other editing capabilities on their host systems, they are terrifically useful. Nonetheless, there is no question that you may prefer to use your own editor to edit MOO texts and verbs. To find out how, please read the help on '@editoptions', '@program' and 'eval'. Using cut and paste from your favorite editor to the MOO is a popular solution. Some MOO clients help with this process as well.

10.0 Conclusion

The two of us have been working together in the MOO environment since the beginnings of 1994 and have by now come to take many of its remarkable powers for granted. Every day, the MOO enables us to sit down with our friends and colleagues from around the world to collaborate on various projects, share expertise, or just talk, all within the context of a lively, grounded community. This chapter is just one small example of what two people nine time zones apart can do with only the MOO for real-time communication. Without even so much as one phone call or face-to-face meeting as a supplement, we think we are fairly safe in concluding that it is indeed possible to MOO without making a sound . . . providing that one does not count all the laughter that goes on at either end of our MOO connection.

We hope this guide will help you to step confidently into a MOO, and that you will find your experiences there as rewarding as ours have been.

Note

1. At the time of this writing, the following MOOs were part of the GNA Inter-MOO Network.

MOO Name	Telnet Address	Port
ATHEMOO	moo.hawaii.edu	9999
AtlantisMOO	atlantis.fe.up.pt	7777
BayMOO	baymoo.org	8888
BioMoo	bioinfo.weizmann.ac.il	8888
CollegeTown	galaxy.bvu.edu	7777

CPDEE MOO	moo.cpdee.ufmg.br	7777
DaMOO	lrc.csun.edu	7777
Diversity University	moo.du.org	8888
GNA-Lab	gnalab.uva.nl	7777
GrassRoots	rdz.stjohns.edu	8888
Internet Public Library	moo.ipl.org	8888
Lingua MOO	lingua.utdallas.edu	8888
MediaMOO	mediamoo.media.mit.edu	8888
Meridian	sky.bellcore.com	7777
MOOsaico	moo.di.uminho.pt	7777
MooWP	it.uwp.edu	7777
Open Forum	homer.ic.gc.ca	8888
PostModernCulture-2	hero.village.virginia.edu	7777
Sci-Fi	moo.ufsm.br	7777
TecfaMOO	tecfamoo.unige.ch	7777

Works Cited

Curtis, Pavel. *The LambdaMOO Programmer's Manual.* 1996. Available at http://ftp.lambda.moo.mud.org/pub/MOO/.

4

MOO Educational Tools

Ken Schweller

In this chapter we will introduce a set of educational tools that simplify the task of holding classes online in a MOO. These tools address the three major technical difficulties in online MOO teaching—maintaining conversational coherence, making resources available, and managing presentations. To maintain conversational coherence and give everyone a fair chance to speak and be heard, we will focus on the $classroom tool with its set of built-in speech moderation controls. To illustrate how academic resources can be made available, we will introduce the blackboard, bulletin board, mailing-list, note, web–MOO client, camera, TV, and robot tools. For managing presentations we will discuss the @paste command, the lecture, slide projector, and web_projector tools, and the use of special classrooms such as Libraries and Stat Labs. Most of the tools we will be discussing were developed by the author for use on MediaMOO, CollegeTown MOO, or Diversity University MOO and are available on our *High Wired* enCore educational MOO core database. We begin our discussion with a look at the serious problem of maintaining conversational coherence and its implications for MOO teaching and learning.

Maintaining Conversational Coherence

Maintaining conversational coherence is perhaps the biggest single challenge in holding an online MOO class. It is difficult to keep conversations and discussions on track when more than three or four people are gathered in the same room and engaged in animate conversation. Under these conditions competing conversational threads emerge continuously, often transforming the screen into a tangled web of interlaced statements, comments, questions, and rejoinders. Although experienced MOO users quickly learn to disentangle the competing threads and even carry on multiple concurrent conversations, the jumbled and quickly scrolling screen can be quite disconcerting to novice users who may feel they have somehow arrived at the cyberspace equivalent of the Tower of Babel. While this can be a quite entertaining party atmosphere, it is not a conducive environment for the kind of serious discussion, presentations, and debates that are integral aspects of any classroom experience.

The difficulty in maintaining conversational coherence in MOO conversations has a lot to do with the nature of the online medium itself. Although it is customary to refer to MOO conversations as being synchronous, in fact there remains a significant amount of asynchronicity in the process. Conversational participants are usually connected to the server on lines that vary considerably in speed, and users on slow connections will always lag behind in the conversation, and their contributions will arrive late and out of context. To the degree these contributions do not merge gracefully with the slipstream of the current thread, they constitute confusing interruptions. This difficulty is aggravated by the fact that MOO speech is not "streamed." Everyone speaks in complete sentences; we do not see another's speech as it is formed word by word. We cannot alter our own contributions as we stay attuned to the emerging words of our conversational partners, nor can we interrupt a person in midspeech. The worst instance of this occurs when a person is connected to a server over raw Telnet and finds that while typing a response others' words either intermix with his or her own partially formed sentence or the room goes "silent" and ongoing conversations cannot be monitored.

A final factor contributing to the difficulty of maintaining conversational coherence is the lack of extralinguistic cues on the MOO for keeping conversations well ordered. There is no easily implemented turn-taking mechanism. In real life if person A and person B are about to speak at the same time, one or the other usually defers. On a MOO a person's intention to speak is not signaled; both persons may speak at once, and since it is unlikely they will express exactly the same opinion, a fork, or divergent conversational thread, is likely to be introduced into the conversation. As listeners respond to one branch of the fork or the other, more and more forks are created. Given these inherent problems, the challenge of holding an online class is to somehow minimize this forking and reestablish coherence so that discussions, presentations, and debates can be carried out smoothly. A good tool for this is the $classroom object.

Creating Classrooms

One of the first things to do if you plan to hold a virtual MOO class is to build yourself a classroom. Having a classroom not only affords you the tools for moderating conversation, but it provides the amenities you expect in a real-life classroom, such as desks, tables, shelves, a clock, a blackboard, and a bulletin board. If your MOO is a well-connected place, you already have a good idea where you would like to build your classroom. Let's imagine that you are in a building called Science Hall and you wish to add a math classroom there. You will use the @dig command as follows to create the new room:

```
@dig west,w to Math Classroom
```

After creating the new room and typing "west" to go there you will find yourself in the new math classroom, where "you see nothing special." Before you forget, be sure to dig an exit back to Science Hall. Your next step will be to convert your room to a specialized $classroom object using the command:

```
@chparent here to $classroom
```

After typing 'look here' you will see the room's new description:

```
You see a rather empty looking classroom. There's a
  blackboard on the front wall and a clock above it.
  Type 'help here' or '@tutorial' for assistance.
You see Teacher's Desk, Big Table, and Bulletin Board.
```

The desk, table, and bulletin board are default furniture items that were added to your room automatically when you converted it to a $classroom. You can keep or delete these items as you see fit. Let's explore some of the room's new features. To see the local time, type 'look clock'. You will see something like:

```
The clock on the wall reads: 9:45 a.m.
The time is: Thu Jun 6 09:45:22 1998 CDT
Thu Jun 6 13:45:22 1998 GMT
```

If you 'look blackboard' you will see that it is empty. To write a welcome message on the board type:

```
'writeb Welcome to our Math Class'
```

If you then 'look blackboard' you will see:

```
==================== BLACKBOARD ====================
1) Welcome to our Math Class!
====================================================
```

Anyone in the room can look at the blackboard and add or erase lines. Typing 'help here' will list all the available blackboard commands, including instructions for rearranging lines or erasing the entire board at once. One very useful command is 'printb', which creates for the user a personal copy of the blackboard contents in the form of a $note object that can be carried around or saved in some location.

Quite likely you will wish to customize your classroom differently from the default setup. Let's say we want to get rid of the default furniture and build our own. Typing '@rmfurn teacher's desk' gets rid of the desk, and similar commands remove the table and bulletin board. Let's start off by creating a new oak desk. Type '@addfurn Oak Desk'. You will then be asked to furnish a description of the desk and whether or not it is sittable. While you are at it, why not "@addfurn Big Table" and "@addfurn Little Table" to give members of your group somewhere to sit. Folks may then "sit Big Table" or "sit Little Table" as they choose.

There are several advantages to providing several sittable objects like couches, desks, tables, and rugs in your classroom. The room is designed to allow differing degrees of conversational moderation and control based on the seating patterns of the room occupants. Folks seated at the same table can converse easily with one another using 'say' commands, but they will only be heard by folks sitting at the same table. This is very important when there are dozens of folks in the room to avoid the confusion of multiple intermingled conversations.

If a person wishes to be heard by the entire room, the person need only type 'speakup I have a question.' and everyone in the room hears

```
Ken speaks up, "I have a question."
```

Other room controls may be toggled on or off to control whether emoting is local to a table or global to the entire room. An immediate benefit of this arrangement is that folks can break up into small discussion groups in the same room. They can still see each other using 'look' but conversations are confined to their table. Of course a person can 'stand' at anytime and be heard by everyone. A person might do this when giving a class presentation or talk so as not to be distracted by extraneous chatter going on at the tables. Use of the 'speakup' command allows an easy interchange of questions and answers.

Any object created using @addfurn can be used as a location for depositing other objects. If you had a notebook with you, you could 'put notebook on Big Table'. Your notebook would then only be visible to people sitting at that table or people standing about who might 'look big table'. This provides an easy way to organize all the objects that tend to accumulate in often-used rooms. An obvious piece of furniture to add to your room at this point is a bookshelf. It should be made nonsittable (unless you don't mind folks sitting on the shelves) and can be used to store dozens of book objects or knickknacks to keep them from cluttering the floor. To get an object from the shelf type 'get book from bookshelf' or 'put object on bookshelf'. Creating walls and bulletin boards provides additional handy places to hang calendars, pictures, and sign-up sheets.

You may open or close the door to your classroom with the command 'door open' or 'door closed'. When the door is closed, only folks who have been registered for the current class session may enter freely. Each classroom has a "sign" message that is displayed when anyone uses the @who command. If we were to type,

```
@session here is " [in session]"
```

then anyone using the @who command would see:

```
Ken (#123) an hour 3 minutes Math Classroom [in session]
```

It is often useful to use the same classroom for different classes meeting at different times. There are room commands that let you multiconfigure the classroom for two or more different classes and permit you to register particular students for particular classes. Registration is very useful in controlling access to the classroom. If the current class setup is "Math," students registered for that class will be able to enter and leave the classroom even if the door is closed, whereas others would be denied entry. This feature is useful if you often find your class being disturbed by guests or other folks dropping in unannounced for a visit while a student presentation or seminar is in progress. Additional $classroom commands are available by typing 'help here', and '@tutorial' brings up an online manual with detailed command instructions and suggestions.

Using Your New Classroom

Once you have converted your room to a $classroom and added a few desks, tables, and bookshelves, you are ready to experiment with the room's various ways to moderate conversation. A good way to understand how the room works is to invite a friend to join you in the room or for you to connect as a guest on a separate screen so you will have at least two people in the room. The key to using the $classroom is to keep in mind the following principles: A person standing can be heard by all; a person sitting can be heard only by those sitting next to him or her. By partitioning the class into folks standing and folks sitting at different locations, you make it possible to comfortably include a dozen or more people in a single room and maintain conversational coherence.

In a lecture situation or for a student presentation it makes most sense for the presenter to be standing while everyone else is sitting. Even if folks at the table do a little chatting, it will be heard only by their seatmates and not by the presenter or folks at other tables. It is often useful to designate one table as a talker's table and another as a quiet table and let folks choose where they wish to sit. Some folks like to chat and make comments during presentations; others prefer to listen without interruptions or wish to have a spam-free session to log the presentation on their own computers. If the class takes the format of a panel or a debate, then it works best to have all panelists or debaters stand while others sit.

The $classroom also permits you to allow/disallow emoting. Toggling the '@stifle on/off' command lets you control whether people can be heard throughout the room when they clap, hiss, or 'squirm uncomfortably'. Another popular activity in $classrooms is to have small-group discussions. This is easily enabled by creating as many separate tables, desks, rugs or whatever as you have groups. A facilitator or leader who is standing can still speak and give instructions and be heard by all groups. Group leaders can 'speakup' or stand to report their group conclusions when they are ready to be heard by all.

Additional Moderated Rooms

The first online MOO Ph.D. defense was carried out in July 1995 on Lingua MOO in an "auditorium" designed by Jan Rune Holmevik (fig. 1). This room has much of the same moderation functionality as the $classroom and exists as the $moderated_room tool in the High Wired enCore.

When configured for panel sessions, only persons seated in the panel area can be heard throughout the room. The audience on the main floor can still talk and emote among themselves, but they can now address the panel only via the command 'ask'. Questions are sent to the moderator, who in turn can retrieve them and field them to the panel. The 'headset' command can be used to filter out text not coming from the panel in both open and panel format for folks who want to listen to the speakers without being disturbed.

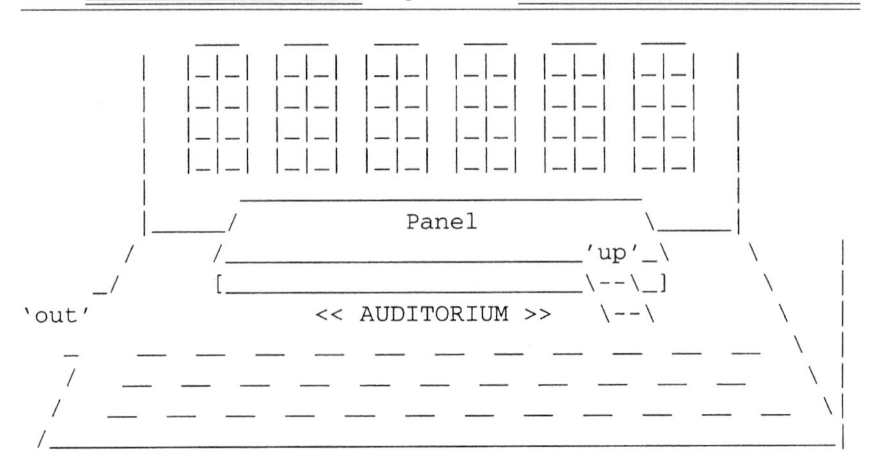

Fig. 1. The Lingua MOO auditorium. (Courtesy of Jan Rune Holmevik.)

Jan has also designed an $intercom tool to monitor and record activity in up to five rooms connected by a central control room. This system was built on behalf of teachers who wished to send their students into different discussion rooms and record their activities. Via the intercom the teacher can also listen in on the discussions and contribute without having to leave the control room. One of the most extensive examples of speech moderation can be seen in the theater complexes on MediaMOO and CollegeTown, where a dozen rooms are linked together to provide fine-grained control over the audio channels for actors, audience, and stage crew.

Classroom Caveats

Some MOO users take the view that conversation-moderating rooms such as the $classroom are inimical to the spirit of online community building. According to this view, imposing conversational control smacks of cyber tyranny and, in the case of environments that purport to be educational, duplicates in this new technology the oppressive mistakes of traditional educational systems. To some extent I agree with this point of view. Transplanting the traditional lecture method wholesale from real-life classrooms to the MOO, for example, is a recipe for disaster and is nearly guaranteed to evoke boredom, irritation, and subtle or open acts of rebellion among class members. MOOs are meant to be places for open participation and conversational exchange on an equal footing. Nevertheless, I feel that proper use and not the misuse of tools such as the $classroom can be liberating rather than oppressive. There is nothing so annoying as trying to follow a speaker's online argument or carry on a serious discussion and being constantly interrupted by extraneous emoting or a bystander's off-topic conversation. Moderation rooms such as the $classroom offer a way to control this conversational confusion and empower the users to dynamically select moderation levels appropriate to a room's changing activities.

Mail Lists and Notes

Once you have a classroom you will need to make resource material available to your group. Let's begin with syllabi, reading lists, and other forms of relatively short text documents. An easy way to distribute short texts is to use the MOO's built-in mail services. Short documents can be MOO-mailed to group members in much the same way you would send documents using regular email. Group members in turn can use MOOmail to submit work or correspond with you asynchronously about class assignments. If your correspondence is frequent, it makes sense to create your own in-MOO mail list. You and your group can read and post messages to and from this list, eliminating the need for posting multiple copies. The list can be configured as private or public, and you may specify who has read/write privileges on it. For help on setting up these lists type 'help mail-system' or consult your MOO administrator.

For class handouts you can use the $note object. Just

```
@create $note named Handout
```

and then '@edit handout' to compose your document. The handout can be dropped in the classroom or on a table and students can all 'read handout'. The classroom blackboard is another easy way to post short information for the students to read when they arrive in class. It is a handy place to post assignments and due dates.

Using the World Wide Web from the MOO

Often the information you wish to make available to your group is already stored at a convenient World Wide Web site, and you would like to provide access to the site from inside the MOO. This is possible using a client such as the Java-based Surf_and_Turf web-MOO client I have developed. To see how this works just point your web browser to http://www.bvu.

edu/ctown/stclient.html and click on the Surf_and_Turf Web-MOO Client button. Your browser screen will split in two. On the bottom half will be the usual MOO login page waiting for you to enter your user name and password. The top portion will be blank until you view some object on the MOO that has an associated web address. It is easy to create such a web-viewable object. Let's say you have a handout in the form of a web page at http://www.myschool.edu/handout.html that you would like your group members to study. Just type:

```
@create $thing named Handout
connect Handout to http://www.myschool.edu/
  handout.html
```

Your handout is now ready for viewing. Typing 'view handout' loads the fully graphical web page into the upper portion of your screen. To share the web page with others just type:

```
display Handout
```

and everyone else in the room will also see the web page displayed on their own browsers. Another tool for displaying web pages is the $webprojector. After @creating a "webprojector" of your own you can preload a list of URL addresses and display them to others in the room one at a time using the command:

```
display 4 on webprojector
```

As usual, assistance in developing web slides is available using the 'help webprojector' command.

The interface of web servers and MOO servers is a very active and ongoing development area, and there are many other varieties of web-MOO interfaces you might wish to explore. Of particular interest are the interfaces developed at BioMOO and in use on Diversity University.

If your system does not support web-MOO clients, it is still possible to access web sites from inside the MOO if you are willing to forgo the graphics and settle for plain text. This is often a good choice anyway if your documents are book chapters or composed of predominantly textual material. In such cases you can use the "web slates" available on most MOOs, which were developed in their original form by Larry Masinter and Erik Ostrom at JaysHouse MOO. You create a web slate by typing:

```
@create $web_slate named Treatise
```

To connect your web slate to the desired site type:

```
goto http://www.myschool.edu/handout.html on
  Treatise
```

You will then see the web page displayed in text format on your MOO screen. You will be able to select hypertext entries by the numbers appearing next to them. For example, you might 'pick 5 on Treatise'. Web slates can be shared. Anyone in the same room can choose to 'watch treatise' along with you. In this way, one person can surf the web while others look on. This is useful when group members wish to work together on web research assignments.

Cameras, Tapes, VCRs, and TV Sets

The $camera object is a flexible tool that permits the "filming" of online events and their replay at a later time. In the context of a text-based environment, *filming* means to record everything in a room that a person in that room might see or hear. It is a transcription of every line that scrolls down the screen from the time the camera is started to the time it is turned off. A common use of this tool is to film ongoing meetings or presentations for later review, but it can also be used to log conversations or official meetings, record interviews, create tutorials, or make movies. This is the tool that was used a few years ago in a film fest on MediaMOO involving about a dozen artistic, dramatic, and documentary film entries. To create a camera and a recording tape type:

```
@create $camera named Camera
@create $tape named Tape1
```

To begin your filming just 'drop Camera', 'put Tape1 in Camera', and 'start Camera'. Everything anyone says or emotes in the presence of the camera from this point until you type 'stop Camera' will be recorded.

Once filming is complete the tape can be viewed using a VCR and TV set. Create these objects in the usual way from $VCR and $TV, then 'hookup VCR to TV', 'put Tape1 in VCR', 'rewind VCR', and 'play VCR'. You will see scrolling down your screen an exact log of the recorded events. Once a tape has been made it can also be added to a "TapeLibrary," which makes it available for viewing by any persons owning TV sets without their having to actually have the tape in their possession. Typing 'schedule TV' gives you a list of tapes that have been made available for viewing in this way. Using this method, folks who were unable to attend a MOO event can watch a tape of the event later at their leisure.

Cameras and tapes are an excellent way to create tutorials. The first tutorial I ever created was called "Coding with Ken" and attempted to teach players how to learn MOO coding. Here is the beginning of what you would see if you chose to 'play 1 on TV':

```
[on TV]     * * * * * * * * * *
[on TV]     COLLEGETOWN - TV Presents
[on TV]     Coding with Ken!
[on TV]     A MOO Programming Tutorial
[on TV]     for Absolute Beginners
[on TV]     Part 1
[on TV]     * * * * * * * * * *

[on TV]     Ken says, "Hi! I'm Ken. I'm here to help
            you learn to program in MOO! This tutorial
            is for Absolute Beginners, no previous
            programming experience in any language is
            assumed!"
[on TV]     Ken says, "MOO, despite all the Bovine
            jokes you may have heard or made yourself,
            actually stands for 'Multi-user Object
            Oriented' programming language."
```

There is about a five-second (programmable) break between each TV event. The tutorial is interactive in the sense that as instructions are given on coding, the viewers are invited to 'pause TV' and actually carry out the instructions as we go. Persons who successfully complete this particular tape will have learned how to program a box of donuts in MOO code.

Tutorials can be created for virtually any subject matter. Folks in the same room can all watch the tutorials on the same TV and chat as they watch. Tutorials can be created by finding a quiet room and just talking in front of a running camera or employing copy/paste techniques similar to those used in creating lectures or slides for closer control of the content.

Conversational Robots

Although few of us in real life have ever had any dealings with a talking robot, this is not an unusual event in MOO communities. Building "bots" is a flourishing enterprise, and robots of all description exist in virtual worlds. These bots vary in their sophistication, but most are somewhat cognizant of their surroundings and able to carry on limited conversations. These bots can be bent to educational uses and with a little programming can do reasonable service as interactive tutors and guides. We will discuss here a bot that was designed especially for educational uses, the so-called conversational robot.

The conversational robot was designed to interact in vigorous conversation with other folks in its vicinity. It has key words and sentence patterns it recognizes and responds to, random responses, and question responses—all user programmable. It has some ability to recognize where it is and to whom it is talking. It was originally designed as a toy, but has since seen serious educational use as a tutor, room guide, information assistant, and "drill and kill" vocabulary tester.

As an example of how it might be used, let us imagine that you are creating a computer museum on the MOO and want to share information on the history of computing with visitors to the room. You might wish to create an Allan Turing robot, in honor of the British mathematician who developed the first theories of computation, and program it to interact with visitors as we imagine Turing himself might have done. To create the bot type:

```
@create $bot named TuringBot
```

To start up the bot, just 'drop TuringBot' and 'activate TuringBot'. To teach the bot to respond to the word *computer,* type 'addword TuringBot' and follow the directions for entering the keyword and appropriate response. Additional instructions for teaching the bot to respond to patterns and questions are available by typing 'help TuringBot'. Here is an example of a conversation I taped with a TuringBot programmed to respond to the words *think, computer,* and *test.*

```
Ken activates TuringBot.
TuringBot says, "Gee thanks for waking me up! I must
  have dozed off."
Ken says, "Can a computer think?"
TuringBot says, "One thing to keep in mind is that a
  computer is anything that can carry out the minimum
  requirements of computation--no matter what it's
  made of."
Ken says, "But to think don't you have to be human?"
```

```
TuringBot says, "I have proposed a test to determine
  whether a machine can truly 'think.'"
Ken says, "what test?"
TuringBot [to Ken]: I introduced the so-called
  Turing's Test of machine intelligence in 1950 in a
  paper titled "Computing Machinery and
  Intelligence."
```

Of course I do not want to leave the impression that all conversations with suitably programmed robots are this coherent. Quite the contrary. To illustrate some of the problems in using these bots here is the rest of the actual transcription.

```
Ken says, "A machine is just a collection of circuits."
TuringBot says, "What is so just a collection of
  circuits about machine?"
Ken says, "I want to know if machines are intelligent."
TuringBot says, "Um, I get the feeling this
  conversation is not going anywhere."
Ken says, "You are mistaken."
TuringBot [to Ken]: Why do you think I am mistaken?
Ken says, "Because a machine is not alive."
TuringBot says, "Suppose machine were not not alive?
  What then?"
```

One of my favorite uses of these bots has been with students in my artificial-intelligence class. Each semester we hold a Turing's Contest in specially prepared rooms on CollegeTown MOO. Each student programs a bot to be conversant in a particular area, the Simpson trial, for example. Not surprisingly, none of the bots has ever come close to fooling a human interrogator for more than a minute or two. The failure illustrates the great complexity of human thought and language and makes clear the difficulty of attempting to program intelligence into a computer.

Using the @Paste Command

Now that we have a meeting place and have examined methods for making source material available to members of our group we will look at how to manage classroom presentations. One of the simplest ways for one person to present information to a group of others is to use the @paste command first developed at JaysHouse MOO and widely available on most MOOs. Suppose you have two windows open on your computer screen, one open to the MOO, one open to a text file that you wish to present to your group members. Use your machine's copy command to capture the text material you wish to display. Move to your MOO window and type '@paste'. Then use your machine's paste command to enter your text and finish by typing in a single period on a line by itself. Everyone in the room will then see something like this:

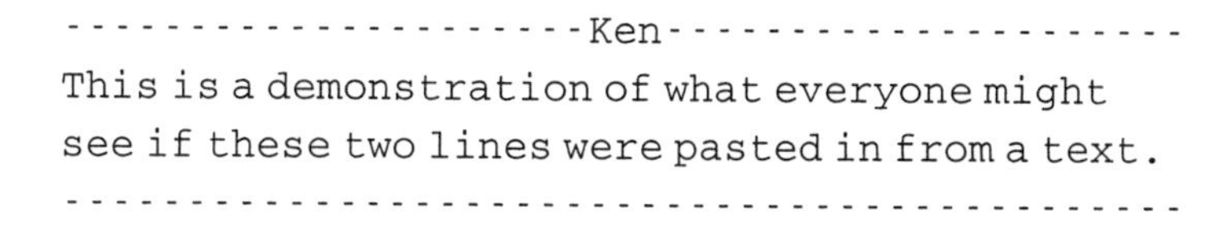

```
--------------------Ken--------------------
This is a demonstration of what everyone might
see if these two lines were pasted in from a text.
-------------------------------------------
```

If your presentation involves more than a few text excerpts, it will be easier to use the lecture tool that we discuss next.

The Lecture Tool

Occasionally you or a member of your group will wish to present material in the form of a lecture. The $lecture tool is especially useful if you have most of the text of your talk already created in a file, and you just wish to have a convenient way to present it on the MOO in small chunks. You certainly do not wish to type in the whole talk as you go using "say" commands. This is much too tedious, slow, and error prone. The $lecture tool allows you to prepare your lines in advance and deliver them at whatever pace and in whichever order you prefer. To create your lecture type:

```
@create $lecture named MyTalk
```

Typing 'help mytalk' will give you a summary of the available commands for creating and presenting your talk. Let's go through the basics. First you must divide the talk into delivery chunks of a few sentences each according to how much you wish to say each time you speak. Generally chunks of one or two sentences are easiest for your listeners to assimilate at one time. Open your favorite off-MOO editor and prepare to port your talk to the MOO by dividing it into chunks separated by blank lines. Next add the word *say* as the first word of each chunk to mark the chunk's beginning; the blank line will delimit the end of the chunk. Here is an example of two lines properly prepared for porting.

```
say Good morning, folks. I am happy to be your guest
  speaker today.
say Thank you for coming blah blah blah
```

Now you are ready to port your talk. Use your editor to copy the entire file. Connect to your MOO and type '@edit mytalk.text'. Then simply paste your file. Finish off by typing 'save' and 'q' to save the file and quit the noteditor. Voila! You have created the lecture and are now ready to test it.

You may 'read mytalk' to preview what you have written. No one else will see what you are reading. You will notice when you do this that each chunk is numbered. Choose which chunk you wish to deliver first and then 'give <chunk number> on mytalk'. You may give the chunks in any order you wish and pause as long as you wish for comments or questions between each chunk delivery. You can repeat chunks for clarity or review as desired. To use the lecture you must be holding it. Anyone holding the lecture has permission to edit it, making the generic lecture an easy way to collaborate with a few others on a shared presentation. Lectures work exceptionally well when combined with graphic web access through a client such as Surf_and_Turf or by having a web browser open concurrently with the MOO server as described earlier. The lecturer can speak in the MOO window while showing slides in the web window. This technique has been successfully used on CollegeTown on several occasions by a group of Alzheimer researchers meeting once a month. The speaker lectured on some aspect of neurophysiology while the session participants simultaneously examined charts and color slides of neurons in their associated web windows.

Using a Slide Projector

A very useful presentation tool is the $slide_projector. This is most useful for displaying tables, lists, charts, and ASCII graphics. You can create a slide projector and a set of slides using

```
@create $slide_projector named Projector
```

You can load the slides with whatever ASCII material you wish up to twenty lines or so and project the slides in any order to be seen by everyone in the room. Figure 2 is a slide showing a picture of my office. The command for showing slides is very similar to the lecture commands. If the ASCII drawing of my room were on slide 10, I would type 'show 10 on projector'.

```
* * * * * * * * * * * * * * * * * * * *
       _________________________________
       |file|       |blackboard|          |
       |cab_|   ___________________       |
       |       ~|               __ |~     |
      /        ~|       rug      s||~     |
      |w       ~|____            o||~     |
      |i       ~|bean\           f||~     |
      |n       ~|_bag/          _a||~     |
      |d        |__________________|~
      |o      ______                    out
      |w     |      |
      \    [ |desk|          KEN'S        |
       |     |____|          OFFICE       |
       |                                  |
       |            |--Book Shelf--|      |
       |_|clock|____|______________|______|
* * * * * * * * * * * * * * * * * * * *
```

Fig. 2. Ken Schweller's CollegeTown office

The creation of slides is once again a matter of copying and pasting from an external file (or building the slides by hand in the MOO using @notedit). A complete tutorial comes in the set of slides packaged with each new projector. Viewing the tutorial slides will instruct you in all the fine details.

Often it makes the most sense on a MOO to create special rooms for special events and lock the appropriate tools and objects inside that room to insure their availability. We conclude this chapter with a look at two such special rooms that you might wish to build on your own MOO.

The Library

If your course is heavily dependent on readings and articles, you might find it advisable to build a separate library room. Libraries consist mainly of books, shelves, card catalogs, and perhaps a robot librarian to assist browsers. Shelves can be created by making your library room an instance of the $classroom and using the '@mkfurn' command. To create MOO books, just create appropriately named web slates, connect them to the appropriate URLs and @describe them with a phrase such as "You see a leather-bound book open to Jane's Treatise on Whatever," or use the 'con-

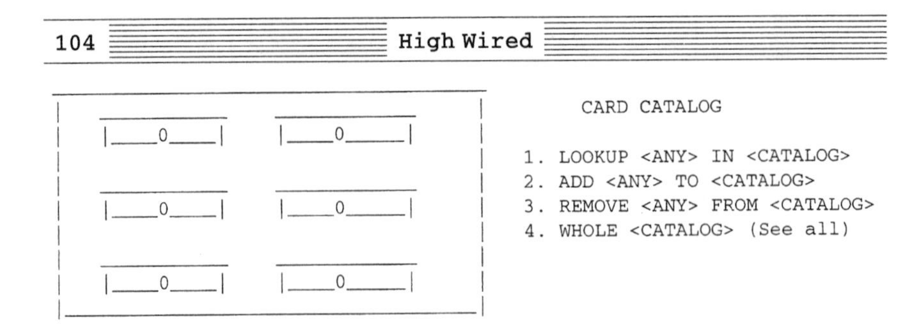

Fig. 3. Card catalog

nect' command to create books viewable by folks on clients similar to Surf _and_Turf. Collections of these books can be created and stored on the library shelves for ready access. Web slate books use storage space efficiently since the only things they actually contain are pointers to text material stored on external servers. To aid in keeping track of your book collection you can create a catalog by typing

```
@create $catalog named Card Catalog
```

A peek at your new card catalog will reveal the possible user commands (fig. 3).

The Lab

An example of a laboratory room is the Probability and Statistics Lab on CollegeTown. This room contains tools for doing simple probability experiments and doing statistical analyses on simple data sets. In this room, for example, you find a sampling urn (fig. 4). If you carry out the experiment suggested in the description, you might see

```
red_Guest picks 10 marbles from the Sampling Urn . . .
Marbles removed: 2 black     8 white
Do you wish to return the marbles to the urn? [Enter
  'yes' or 'no']
```

The urn and various other objects in the room such as coins, dice, and a random-number machine permit the generation of a wide variety of data sets for testing probability hypotheses. Also in the room is a Stats Calculator (fig. 5). This calculator performs some elementary probability and statistical computations and displays the results on the room's projection

You see a jewel-encrusted Sampling Urn that can store any number of black marbles and white marbles for use in probability experiments.

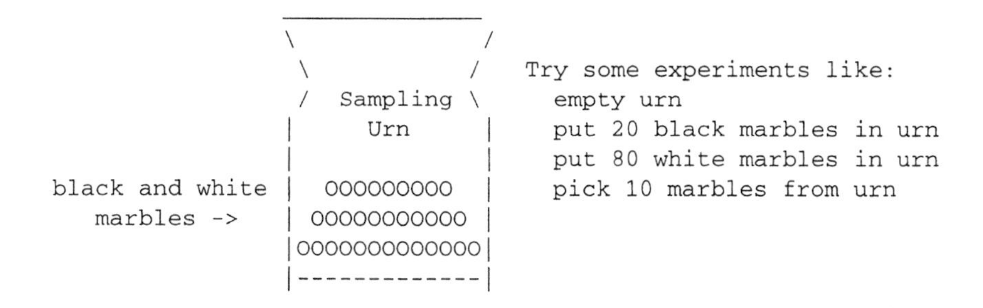
Sampling
Urn
black and white
marbles ->
Try some experiments like:
empty urn
put 20 black marbles in urn
put 80 white marbles in urn
pick 10 marbles from urn

In the urn you see 0 black marbles and 0 white marbles . . .

Fig. 4. Sampling Urn

```
          Stats Calculator
 _______________________________
|    --------------------       |
|    |            120     |     |   Example of use:
|    |____________________|     |    To find the number of combinations
|                               |    of 10 objects taken 3 at a time..
|  Permutations button      (1) |       'press 2 on calculator'
|  Combinations button      (2) |       enter 10
|  Factorial button         (3) |       enter  3
|  Mean/Variance button     (4) |    Answer: 120
|_______________________________|
```

Fig. 5. Stats Calculator

screen where everyone in the room can look at them. To keep folks from walking off with the tools, the tools can be locked to the room. Rooms like this are fun for doing online collaborative experiments in real time and represent the kinds of specialty rooms that give a MOO its unique look and feel.

Conclusion

The use of the $classroom tool and the special resource and presentation tools makes running a class in virtual space, if not always easy, at least much more manageable. I would like to conclude by listing what I consider to be the most important characteristics of a good MOO educational tool. A good tool must be

1. inviting
2. easy to use
3. fun
4. powerful
5. adaptable
6. evocative

One thing that makes a tool inviting and easy to use is its degree of functional similarity to a familiar real-world object. If the commands for virtual TV sets correspond to the actions we perform on real-life TVs, then in some sense we already know how to use these tools, and we adopt them easily. Commands should be easily guessable, and the tool's behavior should be as predictable as possible. This is one reason for building rather old-fashioned looking objects such as blackboards and slide projectors on MOOs. While it would be just as easy to build a TV set that you could walk into or that morphed into a jigsaw puzzle, such an unfamiliar object is more threatening and difficult to assimilate, especially for novice users.

If you can program a bit of whimsy into your tools, all the better; tools that are fun to use will be used often. The best tools must also be simple enough for use by beginners and powerful enough for experienced users. They must scale up from toy demo uses to full-fledged conference use. A slide projector should not break down if it is called into use in a room with three dozen people. Good tools must be adaptable; they should find use in a wide variety of situations and serve as generic patterns for even more specialized tools. Finally, a good tool must be evocative. Just using the tool should quicken the user's imagination to further possible uses. Tools with these characteristics will have a long and active MOO life and be instantiated a thousand times.

5

Taking the MOO by the Horns

How to Design, Set Up, and Manage an Educational MOO

Jan Rune Holmevik and Mark Blanchard

"I am Oz, the Great and Terrible. Why do you seek me?"
Dorothy asked, "Where are you?"
"I am everywhere," answered the Voice, "but to the eyes of common mortals I am invisible."

Although there are public MOOs that provide free space for educators, researchers, and students, there are many good reasons why people want to set up their own MOO. Running your own MOO gives you freedom with regard to choice of theme, purpose, and overall design. Bringing students online and allowing them to help design and build the MOO as well as using it productively in your teaching may not be possible unless you run your own. In addition, setting up a MOO is a fun and rewarding task that will provide you and your university or institution with a unique environment for collaboration, where the only boundaries are your own and your users' imagination and creativity.

In this chapter we will cover the basics of what a new MOO administrator needs to know in order to start an educational MOO. Throughout the chapter we discuss specifically the LambdaMOO server and database, in addition to the *High Wired* enCore, a database that we have designed specifically for educational use. We also briefly mention some of the other systems that are currently available, discuss various machine platforms that are suitable for running a MOO, and describe the kind of permissions the new administrator needs to obtain in order to run a MOO at his or her university or institution.

In the second and main part of the chapter we take you step by step through the process of downloading the software, installing it on your machine, getting the MOO up and running, and customizing the database for your particular machine and Internet environment. Also included in the second part is a discussion of the object-oriented principles of MOO, and the anatomy of the LambdaCore database.

As a MOO administrator there are many factors to consider both technically and otherwise, and the third part of this chapter is devoted to issues that need to be addressed after the MOO is up and running. Among other things, the new administrator must learn how to create new user accounts. This and other issues will be discussed and explained in detail. We also look at some things to consider regarding design of the MOO and its public spaces. With time the administrator may also want to import objects and tools created on other MOOs, and for this reason we include a brief discussion about how that this can be accomplished. Managing a MOO with hundreds of users is a demanding task, and one of the immediate things the administrator must think about is how to compose a team of coadministrators to deal with the various challenges of day-to-day administration. We discuss various ways to do this and what needs to be considered. Finally we talk about some of the responsibilities of being a MOO administrator and the ethic that needs to be observed.

So, You Want Your Own MOO?

If you have not used a MOO before, or if you have only used one sporadically, we strongly recommend that you spend as much time as possible getting to know the MOO system before you start working on setting up your own. Preferably you should have visited several MOOs and become familiar and comfortable with the textual environment and the most-used commands. Chapter 3 in this book is designed to help people who are new to MOO technology get started using it productively. If you have not read this chapter already, we recommend that you go back and do so before reading on.

Getting to know other MOOs will not only give you valuable firsthand experience with how the technology works, but it will also aid you in conceiving and designing your own. The list of MOOs in the appendix of this book is a good place to start. As you venture into cyberspace in your quest to explore and learn more about MOOs, one of the best and most valuable resources you will find is other MOO administrators. Think of MOO technology as an uncharted territory, a frontier where there are few or no road signs telling you where to go or what to do. The people you meet will be your best guides and mentors. No book can teach you everything you need to know in order to become a successful MOO administrator. We can only help you get started and show you some tricks of the trade. In order to create a good MOO for your students and others, first you need to become citizen of MOO world yourself and learn the technology from the inside out.

Unlike most other software programs, a MOO is not a system that you install once and live happily with from upgrade to upgrade. A MOO is a living environment composed of real, creative people. It is a constantly changing, dynamic community that needs attention and management on daily basis. You will get requests for new characters that must be processed, questions of all sorts that need to be answered, social problems that need to be addressed, technical problems that need to be fixed, and so forth. So, in addition to the time it takes to set up the MOO for the first time, which may take anywhere from a few days to several weeks depending on your ambitions and level of expertise, you also need to set aside time for daily administration. How much time you need to spend will again vary depending on the size and purpose of the MOO you want. As a general rule, however, you should plan on an average of thirty minutes to an hour daily online time.

Some Preliminary Design Considerations

It is important that you have an idea of what kind of MOO you want to build before you start the practical work on it. As you will see from the list of MOOs in the appendix to this book, there are a number of different types of MOOs for a number of different purposes. Some educational MOOs are open to the general public; others are not. Some have a clearly defined mission, while others are open to experimentation and exploration. Some are huge, with thousands of registered users from all over the world; others are small and confined to a single university or even department or program. Despite these differences, however, we may still group most educational MOOs as shown in table 4.

The labels "Private" and "Public" denote whether the MOO is open to the general public or not, while the labels "Special" and "General Purpose" say something about the MOO's target audience and purpose. A special-purpose MOO is usually designed with a particular group of people, academic field, or theme in mind, while a general-purpose MOO typically tries to cater to many academic fields, diverse groups of users, and so forth. Because public MOOs are open to the general public and often accept just about anyone who wants a user account there, they tend to grow big very fast. This is especially true if the MOO is also a general-purpose environment. Because of this, these MOOs require a lot of hardware, maintenance, and administration. Private MOOs, on the other hand, especially those that fall in the special-purpose category, are usually much more selective about who they let in the door, and thus they are typically smaller and easier to maintain.

If you want to start out building an open, public MOO that anybody

TABLE 4. A Classification of MOOs

Private, special purpose	Private, general purpose
Public, special purpose	Public, general purpose

can use, you should be aware that this will require considerable resources, both hardware, which we will come to in a bit, and administration. It is our recommendation instead that you start planning for a small, private MOO in which you have a clearly defined academic mission and pedagogical goal. Perhaps an environment designed only for your department, program, or even class. As you gain experience running this smaller MOO, you will later be much better qualified to handle the challenges of operating a large MOO with perhaps thousands of users.

Hardware Considerations

Depending on what type of MOO you envision, you also have to decide what kind of computer system you want to run the MOO on. Generally the best solution, and really the only feasible solution if you plan on building a public MOO, is a dedicated UNIX-based server machine. A dedicated server means simply that you have one machine whose only task is to run the MOO and nothing else. Unless you happen to own such a machine already, or are planning to buy one for your MOO project, the next best solution would be to run the MOO on a shared UNIX server like one of your university's Internet servers. Typically, most university UNIX servers are used for a number of different tasks like running applications, email servers, FTP servers, web servers, and so forth. On a system like this the MOO must compete with all the other programs for processor time, and if the MOO is big, this will certainly lead to a significant slowdown in MOO performance or response time, also known as lag. If you do not have access to a dedicated server for your MOO, running it on a shared server is still a good option. As long as the MOO is fairly small, it would not cause excessive slowdown. Remember also that as your MOO grows bigger you can always move it to dedicated server later.

The LambdaMOO server version 1.8.0, the one we will be discussing in this chapter, will run on several platforms under a UNIX environment. Sun/SPARC, SGI Iris, and DEC Alpha are all ideal host environments for a MOO. Pentium-based PCs and Power Macintosh computers running the free UNIX operating system Linux also make good host machines for a

MOO server and are good and cost-effective alternatives to the workstation systems above.

Regardless of what machine alternative you decide to go with, you need to make sure that the computer you intend to use has a sufficient amount of RAM (memory) installed. As we will explain shortly, a MOO normally keeps all of its data in memory, and for this reason the bigger your MOO is, the more RAM you need. For a small MOO you should plan on at least twenty-four to thirty-two megabytes of RAM, for a medium sized thirty-two to sixty-four megabytes, and for a large MOO as much RAM as you can fit in the machine beyond that. If you plan on using a shared server, these numbers should reflect the amount of free RAM that can be allocated to your MOO. You can always upgrade the amount of RAM in your server as your MOO grows, however, so there is no need to install it all at once.

Permissions

Unless you happen to own or control the machine on which you plan to run the MOO, you need to have some permissions before you start to download the software. Under no circumstance should you set up the MOO at your university's or Internet service provider's (ISP) server without explicit permission. If you do, the system administrators will quickly find out about it, and you may have problems in getting their cooperation later. Running a MOO is not a trivial task for any computer, and it will consume a lot of RAM as well as CPU power on the host machine. So, before you do anything else, be sure that the administrators/owners of the machine on which you plan to run the MOO know what you are doing and that you have their permission to proceed. For many, getting the permission to run a MOO on a university server may prove the hardest part of the whole venture. One reason is that in the past, MUDs and MOOs were often seen as unproductive and time-wasting chat and play spaces and were in fact banned in many universities.

Another thing is that the system administrators may not be happy to have large groups of users from other places come in and use the MOO and thus their machine resources. If your system administrator expresses concern about running a MOO on the university's computers, the plans you made for your MOO and its academic value should be helpful in convincing them that this is something that has been carefully planned and that will have a practical educational purpose. It may also help to enlist support from other faculty members or colleagues in your university or department. This book can be used to demonstrate that educational MOOs are rapidly becoming legitimate and productive learning environments that no school should be without.

Designing and building a MOO should above all be a fun and creative experience, and observing the issues we have mentioned above will help you avoid problems and frustrations and get you the MOO that is right for your particular purpose.

Server and Database

It might be helpful, before we start to download and install the software, to talk a little about the two key components that make up a MOO: the server and the database. It is easy to get these two confused when talking about MOO in general terms, so let us explain what they mean and how they differ from one another.

The server can be thought of as the MOO's operating system. It is the software that handles all the input and output from the MOO, manages user connections, handles the backup routine, and a host of other low-level tasks that the user never sees. There are basically two types of MOO servers available, the original LambdaMOO server by Pavel Curtis, which is the one we will be discussing in this chapter, and the LPMOO server by Robert Leslie. Both servers will run the LambdaMOO database, but apart from that they are quite different. The main difference is that while the LambdaMOO server keeps the entire database in RAM, the LPMOO server keeps only portions of the database in RAM at any given time, while the rest of the database is kept on disk. This means that while the Lambda-MOO server is faster—because reading data from RAM is much faster than reading from a disk—the LPMOO server requires less RAM to run the same database. If you choose to use the LPMOO server rather than the Lambda-MOO server, please check out the LPMOO home page for more information on how to set it up. The URL for this page can be found at the end of this chapter.

The database is often thought of as the MOO itself. From a technical perspective, however, it is nothing more than a large ASCII (American Standard Code for Information Interchange) text file. Storing the database in ASCII format has the great advantage of making it portable, meaning that you can use the database on any computer platform, provided that there is a server for that platform. The database is also often referred to as the *core,* which means that it is the database from which you start to develop your own MOO. We shall return to a more thorough discussion of the core later in this chapter. You have several options with regard to which core database to start with. The two most frequently used are the original

LambdaCore database, developed and maintained by the LambdaMOO community, and the so-called JaysHouse Core, developed by people on JaysHouse MOO, another early and very influential MOO community in cyberspace. In addition, we have developed a special educational MOO core database, the *High Wired* enCore, for use in conjunction with this book. This core is based on the original LambdaCore database and comes with a range of educational tools and features already installed and ready to use. We recommend that you start with this core rather than the original LambdaCore as it will save you a lot of work in developing your MOO. The setup procedure for the *High Wired* enCore is the same as for LambdaCore, detailed later in this chapter. More information about our core can be found on the *High Wired* enCore web site, which is referenced at the end of this chapter.

In addition to the server and database, the word *MOO* also refers to the system's internal programming language. This language is a fully functional object-oriented language that is relatively easy to learn and to use in interesting and productive ways. Over the last five to ten years, object-oriented programming has become the predominant paradigm in computer programming, and the MOO programming language has proven to be a pedagogically very useful tool for teaching students the principles of object-oriented programming and design.

Setting Up a UNIX Directory for Your MOO

Before you start to download the MOO software you need to set up a home directory in which to store all your MOO files. In the following example we will assume that you are setting up the MOO on a machine running SunOS (Solaris), a UNIX operating system from Sun Microsystems. Since most UNIX-based operating systems work the same way and have similar commands, you should be able to use this example even if you are setting up your MOO under Linux or another UNIX-like system. It is beyond the scope of this chapter to teach you everything you need to know about the UNIX operating system, but we will mention some of the commands that you will be using most frequently as a MOO administrator. If you would like to learn more about UNIX, your favorite bookstore should carry several good books on this topic.

When you log into a UNIX machine like your school's Internet server, for example, you get to what is called your home directory. Every user on the system has his or her own directory where their personal files are stored, and this is where we want to put your MOO directory. If you have root access to your local machine, meaning that you are the system administrator of your machine, do not put the MOO in any of the top-level system directories. A MOO in its pure form without any modifications or add-ins is pretty secure, but no program is perfect. It is a much better idea to place the server in an ordinary user account in case potential flaws in the MOO software are found and taken advantage of by someone who wants to break into and harm your system.

The UNIX command for making new directories is *mkdir.* So, in order to make the new directory named "moo" we type:

```
elsie{janruneh}: mkdir moo
```

The construct "elsie{janruneh}" is actually the UNIX prompt on the particular machine we use for the examples in this chapter, so you shouldn't type that. "Elsie" is the name of the machine we are logged on to, and "janruneh" is our user name. On your machine the prompt will look different, but whenever you see the "elsie{janruneh}" prompt in the following examples it means that you should type in what we have written after the colon. As a general rule for all examples in this chapter you should type only what is boldfaced. To show you what happens when we type the various commands, we have included in the examples below some of the system output that our command produces. Unlike most operating systems, UNIX is a case-sensitive system, which means that it distinguishes between small and capital letters. For this reason it is very important that you type the commands exactly as shown.

Now that we have created a new home directory for the MOO software, we go there with the change directory command *cd.*

```
elsie{janruneh}: cd moo
```

To view the contents of any directory we type *dir,* or *ls:*

```
elsie{janruneh}: dir
total 3
gdrwx------   2 janruneh ah     512 May 8 15:51 ./
drwxr--r--    8 janruneh unk   1536 May 8 15:51 ../
```

Downloading the Software

To download the software from the remote server to our local machine we need to use a program called FTP (File Transfer Protocol), which is a UNIX program for transfering files between computers via the Internet. The address to the LambdaMOO server and database archive is ftp.lambda.moo.mud.org, so we FTP there and log on with the user name *anonymous* and use our email address as the password.

```
elsie{janruneh}: ftp ftp.lambda.moo.mud.org
Connected to dulles.placeware.com.
220 dulles FTP server (Version wu-2.4(3) Wed Jun 19
  14:37:15 PDT 1996) ready.
Name (ftp.lambda.moo.mud.org:janruneh): anonymous
331 Guest login ok, send your complete e-mail address
  as password.
Password: me@mysite.edu
230 Guest login ok, access restrictions apply.
```

Just like in UNIX, we can type 'dir' within the FTP program to see the contents of directories. We know that the files we are looking for are stored in the directory /pub/MOO, however, so we type the following ("ftp>" is the prompt so we don't type that):

```
ftp> cd /pub/MOO
250 CWD command successful.
ftp> dir
200 PORT command successful.
150 Opening ASCII mode data connection for /bin/ls.
total 14160
drwxrwxr-x  7 pavel  world     1024 Mar 20 18:43 .
drwxr-xr-x  4 root   nobody     512 Apr 21 21:48 ..
-rw-rw-r--  1 pavel  world   364908 Feb 27 21:42 HelpSystem-02Jun93.txt
-rw-rw-r--  1 pavel  world  2195405 Feb 27 21:37 LambdaCore-02Feb97.db
-rw-rw-r--  1 pavel  world   824030 Feb 27 21:38 LambdaCore-02Feb97.db.Z
lrwxrwxrwx  1 pavel  world       21 Mar 20 18:43 LambdaCore-latest.db
lrwxrwxrwx  1 pavel  world       23 Mar 20 18:43 LambdaCore-latest.db.Z
-rw-rw-r--  1 pavel  world   573973 Mar 20 18:35 LambdaMOO-1.8.0p6.tar.Z
-rw-rw-r--  1 pavel  world   345477 Mar 20 18:36 LambdaMOO-1.8.0p6.tar.gz
lrwxrwxrwx  1 pavel  world       23 Mar 20 18:43 LambdaMOO-latest.tar.Z
lrwxrwxrwx  1 pavel  world       24 Mar 20 18:43 LambdaMOO-latest.tar.gz
-rw-rw-r--  1 pavel  world   351552 Mar 20 18:36 ProgrammersManual.dvi
-rw-rw-r--  1 pavel  world   151354 Mar 20 18:36 ProgrammersManual.dvi.Z
-rw-rw-r--  1 pavel  world   109815 Mar 20 18:36 ProgrammersManual.dvi.gz
-rw-rw-r--  1 pavel  world   265323 Mar 20 18:37 ProgrammersManual.html
-rw-rw-r--  1 pavel  world   246244 Mar 20 18:37 ProgrammersManual.info
-rw-rw-r--  1 pavel  world   488600 Mar 20 18:37 ProgrammersManual.ps
-rw-rw-r--  1 pavel  world   193877 Mar 20 18:38 ProgrammersManual.ps.Z
-rw-rw-r--  1 pavel  world   154857 Mar 20 18:38 ProgrammersManual.ps.gz
-rw-rw-r--  1 pavel  world   251661 Mar 20 18:38 ProgrammersManual.texinfo
-rw-rw-r--  1 pavel  world   222966 Mar 20 18:38 ProgrammersManual.txt
-rw-rw-r--  1 pavel  world     7671 Mar 20 18:38 ProgrammersManual_toc.html
drwxrwxr-x  2 pavel  world      512 Mar 20 18:42 archive
```

```
drwxrwxr-x 3 pavel  world     512 Mar 10 21:40 contrib
drwxrwxr-x 2 pavel  world    3072 Feb 27 22:20 html
-rw-rw-r-- 1 pavel  world  117413 Mar 20 18:41 html.tar.Z
-rw-rw-r-- 1 pavel  world   81520 Mar 20 18:41 html.tar.gz
drwxrwxr-x 5 pavel  world     512 Feb 27 22:13 lambda
drwxrwxr-x 2 pavel  world    1536 Feb 27 22:24 papers
-rw-rw-r-- 1 pavel  world   91742 Feb 27 21:42 texinfo.tex
226 Transfer complete.
2314 bytes received in 0.23 seconds (9.8 Kbytes/s)
```

At the time of this writing, these were the contents of the top level of the LambdaMOO archive. As we can see, there are several files and directories there, but the ones we need are called LambdaMOO-latest.tar.gz, which is the server, and LambdaCore-latest.db, which is the database. Unlike the database, which is nothing more than a huge text file, the server is compressed and stored in binary form (see below). The FTP program needs to be told what format, text or binary, the file is in in order to download it sucessfully. So, in the example below note that we change the download format from ASCII, which is the default, to binary after we have downloaded the database.

```
ftp> get LambdaCore-latest.db
200 PORT command successful.
150 Opening ASCII mode data connection for LambdaCore-
  latest.db (2195405 bytes).
226 Transfer complete.
local: LambdaCore-latest.db remote: LambdaCore-
  latest.db
2447668 bytes received in 3.4e+02 seconds (7.1 Kbytes/s)
ftp> binary
200 Type set to I.
ftp> get LambdaMOO-latest.tar.gz
200 PORT command successful.
150 Opening BINARY mode data connection for LambdaMOO-
  latest.tar.gz (345477 bytes).
226 Transfer complete.
local: LambdaMOO-latest.tar.gz remote: LambdaMOO-
  latest.tar.gz
345477 bytes received in 44 seconds (7.7 Kbytes/s)
ftp> bye
221 Goodbye.
```

To make sure that we got the files we wanted, we can now type 'dir' again to view the contents of our MOO directory.

```
elsie{janruneh}: dir
total 2507
drwx------   2 janruneh ah         512 May 8 15:58 ./
drwxr--r--   8 janruneh unk       1536 May 8 15:51 ../
-rw-------   1 janruneh ah     2195405 May 8 16:04 LambdaCore-latest.db
-rw-------   1 janruneh ah      345477 May 8 15:57 LambdaMOO-latest.tar.gz
```

To save space on the remote server as well as download time, files are usually stored in some kind of compressed format. In UNIX the most commonly used programs for compressing files are called *compress* and *gzip*. A file compressed with compress has the extension *Z*, while a file compressed with gzip has the extention *gz*. Note that the gzip program can also decompress files that have been compressed with compress. Furthermore, before compressing files from a directory with one or more subdirectories, it is common practice to first make one single archive file out of the whole directory structure. The program that is used to make this archive is called *tar*, and archive files carry the extention *tar*. As we can see from the directory listing above, the LambdaMOO server file has been archived with tar and compressed with gzip, so before we can do anything with it we need to decompress and unpack it. The following two commands will accomplish this:

```
elsie{janruneh}: gzip -d LambdaMOO-latest.tar.gz
elsie{janruneh}: tar -xvf LambdaMOO-latest.tar
```

If you want to know more about the gzip and tar commands you can type 'man gzip' and 'man tar' from your UNIX prompt. This will bring up the UNIX manual pages for these two commands. The manual pages cover most of the UNIX commands on your system, so if you run into problems with any of the other commands we use in these examples, please look up the manual page entry for that command.

```
elsie{janruneh}: dir
total 3766
drwx------   3 janruneh ah         512 May 8 16:06 ./
drwxr--r--   8 janruneh unk       1536 May 8 15:51 ../
-rw-------   1 janruneh ah     2195405 May 8 16:04 LambdaCore-latest.db
-rw-------   1 janruneh ah     1638400 May 8 15:57 LambdaMOO-latest.tar
drwx------   3 janruneh ah        3072 Mar 11 14:55 MOO-1.8.0p6/
elsie{janruneh}: rm LambdaMOO-latest.tar
rm: remove LambdaMOO-latest.tar (yes/no)? yes
```

Another look at our MOO directory tells us that the server was successfully decompressed and unarchived. It has been placed in a new, automatically created directory called MOO-1.8.0p6, which we will call the MOO *source directory.* Before we go there to continue work on setting it up, we can delete the original archive file with the *rm* command to save space on our hard disk (see example above).

Compiling the Server

The server is distributed in source code, which means that we must compile it into executable binary form for our particular machine before we can use it. Before we can do this, however, we need to run the *configure* script that can be found in the MOO source directory. This script will take several minutes to complete, and it makes a file called config.h that helps the compilation process. The config.h file holds vital information about the types of networking that the MOO will be using, definitions for the machine that we are using, and so on. If you plan on using the MOO as a public place, you should also edit the options.h file that holds the preferences that are used in the compilation process. The LambdaCore allows you to set email addresses of people in the MOO and also send email to those people. This is very useful when new characters are created and you want the MOO to email them their user name and password automatically. If this is something that you are eventually going to want to do, you should edit the options.h file to open up outbound network connections from the MOO.

In this example we are going to edit the options.h file with the UNIX text editor *vi,* but other UNIX text editors like PICO or emacs can also be used. For those of you who are accustomed to the WYSIWYG editors (What You See Is What You Get) of the Macintosh or Windows environments, vi is probably going to seem awkward and bare. However, its main advantage is that it comes with all UNIX systems, and for our purpose here, it will do the job. Table 5 lists the vi function keys we are going to use in this example.

To start editing the options.h file, we change to the MOO source directory and start the vi program by typing its name followed by the name of the file we wish to edit.

```
elsie{janruneh}: cd MOO-1.8.0p6
elsie{janruneh}: vi options.h
```

TABLE 5. Some vi Commands

Key	Function
i	enter insert mode
esc	exit insert mode
h	move cursor left
l	move cursor right
j	move cursor down one line
k	move cursor up one line
dw	delete word at cursor position
:wq	write changes to file and quit program
:q!	quit vi without saving changes to file

We now see the following text on the screen:

```
/*****************************************************
  Copyright (c) 1992, 1995, 1996 Xerox Corporation. All
  rights reserved. Portions of this code were written by
  Stephen White, aka ghond. Use and copying of this
  software and preparation of derivative works based upon
  this software are permitted. Any distribution of this
  software or derivative works must comply with all
  applicable United States export control laws. This
  software is made available AS IS, and Xerox Corporation
  makes no warranty about the software, its performance or
  its conformity to any specification. Any person
  obtaining a copy of this software is requested to send
  their name and post office or electronic mail address to:
    Pavel Curtis
    Xerox PARC
    3333 Coyote Hill Rd.
    Palo Alto, CA 94304
    Pavel@Xerox.Com
*****************************************************/
```

This is the beginning of the options.h file. In order to enable outbound networking we need to uncomment the following line:

```
/* #define OUTBOUND_NETWORK */
```

To get to this line we press the *l* key (move cursor down) on the keyboard several times until it appears on our screen. As we can see, the line is enclosed by the following set of characters: /* and */ . These character constructs are called comments and are normally used by programmers to annotate or explain the programs they write. When the compiler finds something that is enclosed in comments it discards it and jumps to the next line, and thus it is not included in the final executable program. Therefore, we need to remove the comments on this line in order for compiler to enable outbound networking for our MOO. To do this, we press the keys 'h' (move cursor left) and 'l' (move cursor right) to move the cursor to the comment and then press the 'd' and 'w' keys to delete the comments one at a time. When we have done this, we press the key combination ':wq' to write the changes to disk and quit vi. If you mess up and delete something you should not have, press the key combination ':q!' to quit vi without saving changes. Then go back and repeat the whole process again.

Once we have edited the options.h file, we can run the configure script that we mentioned earlier. To do this simply type 'configure'.

```
elsie{janruneh}: configure
checking for bison
checking for gcc
checking how to run the C preprocessor
checking whether -traditional is needed
checking how to run the C preprocessor
```

The script will take a few minutes to check that our system has everything that is needed to compile the MOO server. Once it is finished, the following text appears on our screen:

```
checking which MOO networking configurations are
  likely to work . . .
checking for sys/socket.h
checking for telnet
-------------------------------------------
| The following networking configurations will
| probably work on your system; any configuration
```

```
| *not* listed here almost certainly will *not* work on
| your system:
|
| NP_SINGLE NS_BSD/NP_LOCAL NS_BSD/NP_TCP NS_SYSV/
  NP_LOCAL NS_SYSV/NP_TCP
-----------------------------------------------------
creating config.status
creating Makefile
creating config.h
```

We are now finally ready to compile the server. This is actually the easiest part of setting up the MOO. All you have to do is type 'make', and the system will take care of the rest. The compilation process takes several minutes and produces an output that looks like this:

```
elsie{janruneh}: make
gcc -Wall -Wwrite-strings -g -O -c ast.c
gcc -Wall -Wwrite-strings -g -O -c code_gen.c
gcc -Wall -Wwrite-strings -g -O -c db_file.c
gcc -Wall -Wwrite-strings -g -O -c db_io.c
gcc -Wall -Wwrite-strings -g -O -c db_objects.c
gcc -Wall -Wwrite-strings -g -O -c db_properties.c
```

Once the compilation process finishes, two new files appear in the MOO source directory, one called moo, which is the MOO server itself, and one called restart, which is a script that we will use to start the MOO with. On some systems these two files will have an asterisk (*) after their name signifying that they are executable binary files. The next thing we do is to copy these two files up to our MOO home directory with the *cp* command. Note that you should substitute for /home/zen/janruneh/moo in the example below the path to the moo directory on your particular machine. Use the pwd command to find the path if you don't know it.

```
elsie{janruneh}: cp moo /home/zen/janruneh/moo
elsie{janruneh}: cp restart /home/zen/janruneh/moo
elsie{janruneh}: cd /home/zen/janruneh/moo
elsie{janruneh}: dir
total 3098
drwx------ 3 janruneh ah        512 May  8 20:18 ./
drwxr--r-- 8 janruneh unk      1536 May  8 20:17 ../
-rw------- 1 janruneh ah    2195405 May  8 16:04 LambdaCore-latest.db
drwx------ 3 janruneh ah       4608 May  8 18:32 MOO-1.8.0p6/
```

```
-rwx------ 1 janruneh ah     949028 May  8 20:18 moo*
-rwx------ 1 janruneh ah       2016 May  8 20:18 restart*
```

We are now almost ready to start our new MOO, but first we want to rename the database file to something that reflects the name of our MOO. In this example we use the name mymoo, but you can substitute this for the name you wish to give your own MOO. The UNIX command for renaming files is *mv <old file name> <new file name>*. In addition, we need to make an extra copy of the database that will be used by the restart script for backup purposes.

```
elsie{janruneh}: mv LambdaCore-latest.db mymoo
elsie{janruneh}: cp mymoo mymoo.db
```

We start the MOO server by typing 'restart mymoo'. By default the MOO server will listen for incoming connections on port 7777, so in order to log on we must Telnet to that port on our local machine. The domain name for the machine in this example is elsie.utdallas.edu. You should use the domain name or IP number for the machine on which you started your MOO when you log on.

```
elsie{janruneh}: restart mymoo
[1] 20581
elsie{janruneh}: telnet elsie.utdallas.edu 7777
Trying 129.110.16.9 . . .
Connected to elsie.utdallas.edu.
Escape character is '^]'.
Welcome to the LambdaCore database.

Type 'connect wizard' to log in.

You will probably want to change this text and the
  output of the 'help' command, which are stored in
  $login.welcome_message and $login.help_message,
  respectively.
connect wizard
*** Connected ***
The First Room
This is all there is right now.
Your previous connection was before we started keeping
  track.
```

```
There is new news. Type 'news' to read all news or 'news
  new' to read just new news.
You say, "Hi and congratulations! You have just logged
  on to your new MOO :-)"
```

The Anatomy of the LambdaCore Database: The System Object, Generic Classes, and Utility Packages

We have mentioned the core of the database several times in this chapter, and it is now time to explain what we mean by that word. One way to explain it is to say that the core is the database you start with when you first set up your MOO. This core database has only one room in it (the First Room; see above), and the only user account is the wizard character that you use to log in with the first time. The core may seem quite bare at first glance, but it does contain everything you need, and as you make new rooms and create new objects, you build your own database in layers around the core. The core of your database is defined on the system object, which for this reason is the most important object in the MOO. It holds information vital to the proper operation of the MOO, and you should be very careful when modifying it because errors will almost certainly result in major malfunctions. Only administrators or wizards can modify the core, so you do not have to worry about accidental damage caused by other players in your MOO. To give you a better idea of what actually constitutes the core we have included a listing of some of the core items that are defined on the system object in LambdaCore, version 02Feb97, which is the one we are using in this example, and which is also the basis for the *High Wired* enCore.

```
The System Object (#0) [readable]
  Child of Root Class (#1).
.builder              Wizard (#2)    r c    #4
.login                Wizard (#2)    r      #10
.limbo                Wizard (#2)    r c    #15
.registration_db      Wizard (#2)    r c    #16
.wiz_utils            Wizard (#2)    r c    #24
.guest                Wizard (#2)    r      #31
.hacker               Wizard (#2)    r c    #36
.player_db            Wizard (#2)    r      #39
.gender_utils         Wizard (#2)    r      #41
```

```
.mail_editor            Wizard (#2)    r c    #47
.note_editor            Wizard (#2)    r c    #48
.verb_editor            Wizard (#2)    r c    #49
.generic_editor         Wizard (#2)    r c    #50
.object_utils           Wizard (#2)    r c    #52
.letter                 Wizard (#2)    r c    #54
.dump_interval          Wizard (#2)    r c    3600
.list_utils             Wizard (#2)    r c    #55
.command_utils          Wizard (#2)    r c    #56
.player                 Wizard (#2)    r c    #6
.wiz                    Wizard (#2)    r c    #57
.prog                   Wizard (#2)    r c    #58
.code_utils             Wizard (#2)    r c    #59
.help                   Wizard (#2)    r c    #60
.nothing                Wizard (#2)    r c    #-1
.building_utils         Wizard (#2)    r c    #21
.string_utils           Wizard (#2)    r c    #20
.news                   Wizard (#2)    r c    #61
.note                   Wizard (#2)    r c    #9
.container              Wizard (#2)    r c    #8
.thing                  Wizard (#2)    r c    #5
.exit                   Wizard (#2)    r c    #7
.room                   Wizard (#2)    r c    #3
.player_start           Wizard (#2)    r c    #62
.root_class             Wizard (#2)    r c    #1
.recycler               Wizard (#2)    r c    #63
.garbage                Wizard (#2)    r c    #64
.housekeeper            Wizard (#2)    r c    #71
.network                Wizard (#2)    r c    #72
.sysobj                 Wizard (#2)    r      #0
.byte_quota_utils       Wizard (#2)    r c    #80
.object_quota_utils     Wizard (#2)    r c    #82
.server_options         Wizard (#2)    r c    #83
```

In the leftmost column are the names of some of the core objects that are defined on the system object in our MOO. The second column tells us who owns these objects, while the last column shows us the number of each object. While objects can share the same name (except player objects, which must always have unique names), they always have their own unique object number that we can use in reference to these objects. For convenience, core objects like the ones listed above can also be referenced

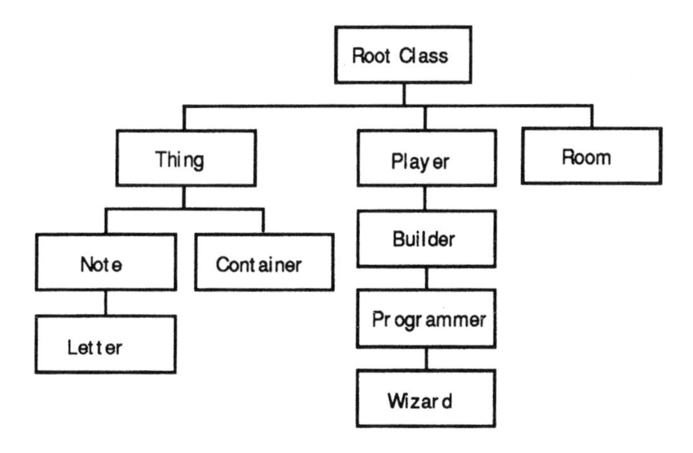

Fig. 6. Simplified overview of the top level LambdaMOO class hierarchy

by putting a dollar sign ($) in front of their names. As a MOO administrator or wizard you can add objects to your core definition at any time by adding their name and object number to the system object as properties.

Earlier in this chapter we said that MOO is an object-oriented system. In short, this means that the database is organized as a hierarchy of *objects* that are interconnected through a mechanism called *inheritance* (see fig. 6). The object is the basic, fundamental unit in every object-oriented system and is composed of two things, programs, in MOO called *verbs,* and data, which in the MOO are called *properties.* Objects that share a common set of verbs and properties through the inheritance mechanism are organized in what are called *classes.* In the LambdaMOO database all objects *descend* from the root class, which has object #1. Under the root class level there are a number of *subclasses* as illustrated in figure 6.

In object-oriented terms we say that the classes *thing, player,* and *room* are subclasses (children) of the root class. The classes *note* and *container* are subclasses of the class *thing,* while the three classes *builder, programmer,* and *wizard* are all subclasses of the class *player.* Because a class object can be used as a template or blueprint for creating new objects and subclasses, we say that they are *generic.* When later in this chapter we explain how to make new characters in the MOO, what we are really doing is creating a new object that is a child of one of the classes player, builder, programmer, or wizard.

Utility packages are another important part of the database core. These packages provide a convenient way to store and reference frequently used verbs and properties that belong together but may or may not be associated with any particular class. An example of this is network utilities package, which we work with in a bit when we start to customize our database. Other important utility packages in the LambdaCore include the wizard utilities that contain important and useful verbs for wizards and the object utilities that contain a set of verbs for manipulating objects and many others.

Customizing the Core

We are almost ready to start building in our new MOO, but first we should customize some preferences in our core. Among other things, we need to set a few network preferences to allow the outbound networking that we enabled before we compiled the server to work properly. In doing this we will be using the two MOO commands @display and @set. The @display command is described in detail in chapter 3, so we will only summarize it briefly here. The @set command has not been mentioned before, so we are going to talk about it in more detail. The LambdaCore database also comes with a very extensive help database that you can use to get additional help on these and other commands.

The command @display (can be shortened to @d) is used to list verbs and properties on an object. To list all properties on the object type

```
@display <object name>.
```

To list all verbs on the object type

```
@display <object name>:
```

Note that the period (.) after the object name tells the MOO to list properties, while the colon (:) tells it to list verbs. We will show you a practical example of how @display is used shortly.

The @set command is used to assign a value (data) to a property. The value that a given property can be assigned depends on its datatype, which in the MOO can be either an object number, an integer, a text string, or a list containing several object numbers, integers, or text strings. This last composite datatype is also called a matrix. For more information on properties and datatypes we must refer you to the *LambdaMOO Programmers Manual* (see references at the end of this chapter). For our purpose here we will show you how the @set command is used to assign a value to a property. Say that we have an object in our MOO called *Computer.* This object has a number of properties, among them the properties "brand" and "model." To assign values to these two properties we type the following:

```
@set computer.brand to "Macintosh"
@set computer.model to "PowerBook 3400c"
```

You will learn more about datatypes and properties as you start to work on your MOO and explore the help system and other online resources. For now, however, this should enable you to set and customize preferences in your core and get on with building your MOO.

The most important group of preferences we must set is found on the network utilities package. Inside the MOO this package can be referenced as $network (see the list of core objects above). In order to list and view them we type:

```
@display $network.
Network Utilities (#72) [ readable ]
  Child of Root Class (#1).
.site                          Wizard (#2)    r     "yoursite"
.large_domains                 Wizard (#2)    r     { . . . }
.connect_connections_to        Wizard (#2)          {}
.postmaster                    Wizard (#2)    r c   "postmastername@yourhost".
.port                          Wizard (#2)    r c   7777
.MOO_name                      Wizard (#2)    r c   "YourMOO"
.valid_host_regexp             Wizard (#2)    r c   "^%([-_a-z0-9]+%.%"+ . . .
.maildrop                      Wizard (#2)    r c   "localhost"
.trusts                        Wizard (#2)    r     {#36}
.reply_address                 Wizard (#2)    r c   "moomailreplyto@yourhost".
.active                        Wizard (#2)    r c   0
.valid_email_regexp            Wizard (#2)    r c   "^[-a-z0-9_!.%+$'="+ . . .
.invalid_userids               Wizard (#2)    r c   {"," "sysadmin," . . . }
.debugging                     Wizard (#2)    r c   0
.errors_to_address             Wizard (#2)    r c   "moomailerrors@yourhost"
.suspicious_userids            Wizard (#2)    r c   {"," "sysadmin," . . . }
.usual_postmaster              Wizard (#2)    r c   "postmastername@yourhost".
.password_postmaster           Wizard (#2)    r c   "postmastername@yourhost".
.queued_mail                   Hacker (#36)         {}
.queued_mail_task              Hacker (#36)   r     970545644
.envelope_from                 Wizard (#2)    r c   "postmastername@yourhost".
.blank_envelope                Wizard (#2)    r c   0
```

As we can see, the preferences in the LambdaCore are all generic values and need to be modified for your particular computer and Internet environment. Below we have listed the properties that you need to change. Type in the following lines and substitute our generic values for your own specific ones.

```
@set $network.site to "<enter the domain name of your site
here>"
```

```
    Example: @set $network.site to "utdallas.edu"
@set $network.postmaster to "<enter your email address
here>"
    Example: @set $network.postmaster to
    "dorothy@utdallas.edu"
@set $network.MOO_name to "<enter the name of your MOO
here>"
    Example: @set $network.MOO_name to "Lingua MOO"
@set $network.maildrop to "<enter the domain name of your
email gateway>"
    Example: @set $network.maildrop to "utdallas.edu"
@set $network.reply_address to "<enter your email
address here>"
@set $network.active to 1
    This will turn on outbound networking and allow you to
    send email from your MOO.
@set $network.errors_to_address to "<enter your email
address here>"
```

The other network preferences can be left as they are for now. You can always go back and change them at any time if you need to. Next we need to change a few of the login preferences. By default, automatic character creation is turned on, which means that anyone who logs on to your MOO as a guest can have the system create a character for them automatically. In an educational MOO we feel that it is important to control who has access to the system, so we want to disable the automatic character creation. If, however, this is a feature you would like to keep, you can skip to the next paragraph. We will not list all the properties on $login here, but if you want to do it in your MOO simply type:

```
@display $login.
```

The property we need to change is called "create_enabeled," and it must be assigned the value 0. To do this we type:

```
@set $login.create_enabled to 0
```

When you first logged on to your MOO you probably saw the generic welcome screen welcoming you to the LambdaCore Database. This is the first thing people see when they log on to your MOO, so you should edit it as soon as possible. You may include any information you like, but to give

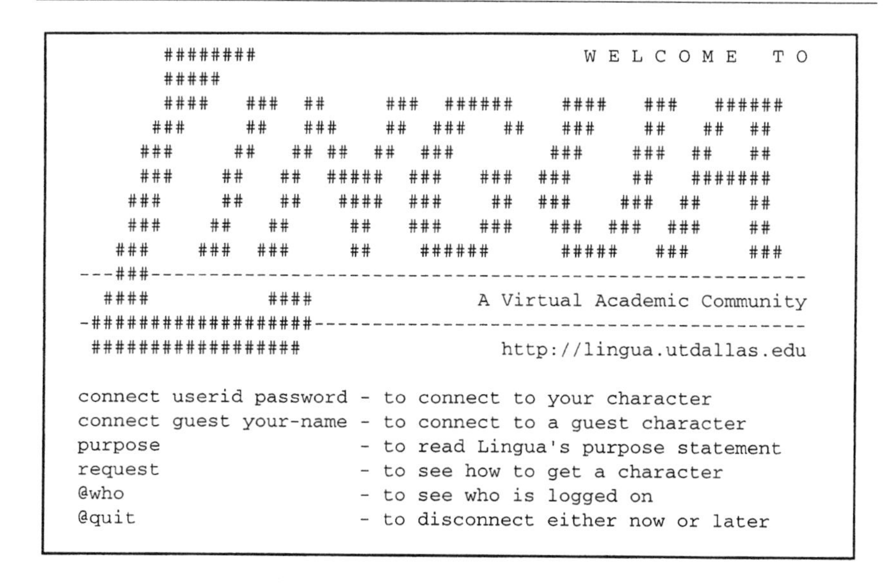

Fig. 7. Lingua MOO login screen

you an example of how a welcome screen might look we have included the one from Lingua MOO (fig. 7).

You can use the MOO's note editor to edit your welcome screen. For more on how to use the note editor and other in-MOO editors we refer you to chapter 3. To invoke the editor and edit the property that holds the welcome screen information you type:

```
@notedit $login.welcome_message
```

We have now covered the most important preferences that should be set before you start to work on designing and building your MOO. There are many more that you can set, in fact so many that we cannot possibly cover them all in this chapter. However, as you get more experienced and get to know the LambdaCore better you can get back to these other options later.

Inhabitants of the MOO World: Players, Builders, Programmers, Wizards, and Guests

In the MOO there are four so-called character types or player classes. These are: Player, Builder, Programmer, and Wizard. The difference between them has to do with the level of access each class has to the MOO's

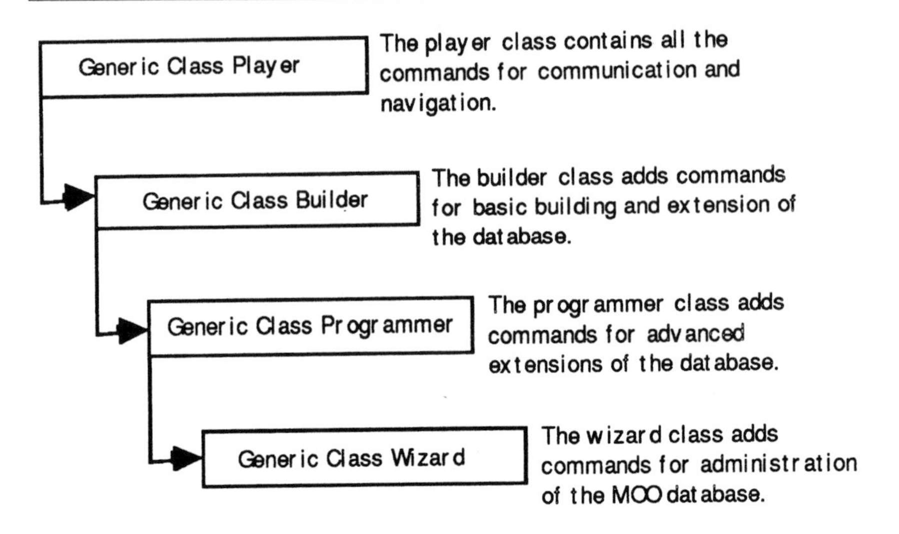

Fig. 8. MOO player class hierarchy

system resources, and the number of commands they have available to them.

As illustrated in figure 8, the generic player class is the basis for the three other player classes and contains all the commands you need in order to talk to other people online, move around in the virtual world, send and receive MOO mail, and much more. *Player* is the default class to which all newly-created characters belong. We will take you through the process of creating new characters shortly, but first let us look at the other three player classes you will be working with in the MOO.

One thing that attracts people to MOO is the fact that it is a user-extendable system, which means that people can help build and shape the virtual world in creative and imaginative ways. What makes this possible are the commands found on the generic builder class like @dig and @create, which were explained in detail in chapter 3. A character belonging to this class will have access to all these building commands, and because of the way inheritance works in an object-oriented system (see "The Anatomy of the LambdaCore Database" above), the character will also have access to all the commands available on the basic player class.

Another exciting and powerful feature that distinguishes a MOO from most other popular MUD systems is the internal programming language that we talked about in the beginning of this chapter. By means of this object-oriented programming language people can add new verbs (programs) to the MOO and thus help change the virtual world in much more profound ways can be accomplished with the basic building tools we mentioned above. All the programming commands are defined on the generic programmer class and are available only to wizards or people who have been given programmer status. As with the builder class above, the principle of inheritance gives programmers access to both builder commands and basic player commands. Unfortunately, there is no room in this chapter to discuss MOO programming in specific, but for those who are interested in learning more about it, Pavel Curtis's *LambdaMOO Programmer's Manual* is a good place to start.

The most powerful class in the MOO is the generic wizard class. This class inherits all the commands available to players, builders, and programmers and has a set of special commands for database maintenance and character administration. We will return to some of the wizard commands that you will be using most frequently toward the end of this chapter.

In addition to these main player types, there is also a special character type called *guest*. This is not really a class in itself, but rather a special reusable player character that people who want to visit the MOO without necessarily wanting a permanent character there can use. As opposed to permanent characters belonging to the four regular player classes, guests are for all practical purposes anonymous, and this poses some special problems that we will return to in more detail.

Creating New Characters for Your MOO

Now that you know a little bit more about the various character types in the MOO we are going to show you an example of how you can create new characters and assign them to the various classes we have discussed.

The wizard character that you used in order to log on for the first time is really a crucial part of the proper function of your MOO and should normally not be used as a regular player character. Thus, before you do anything else, you should create a new wizard character for yourself. There is no command that allows you to create this new wizard character in one step, so in this example we first show you how to make a regular player, then how to make that player a builder, then a programmer, and then finally a wizard.

The command that is always used to create new characters is called '@make-player'. The arguments for this command are the user name for the new character and the email address to the person for which you are creating it. If you have enabled outbound networking for your MOO, which we discussed earlier in this chapter, it will automatically email the user name (which is the same as the login name) and password directly to the person's Internet email address. It is therefore very important that you double-check the email address for accuracy before you press enter and execute the command. If you have not enabeled outbound networking, you must write down the login name and password that the MOO tells you (see example below) and email it to the person yourself. Let us now create a new character named Dorothy with the email address dorothy@wizard.oz. Please substitute the name and email address in this example for the MOO name you want for yourself and your own Internet email address.

@make-player Dorothy dorothy@wizard.oz

The MOO will respond with the following output:

```
Dorothy (#100) created with password 'beSox' for
  dorothy@wizard.oz
Send email to dorothy@wizard.oz with password? [Enter
  'yes' or 'no']
yes
*Player Creation Log (#17) has just been sent new mail
  by Wizard (#2).
Sending the password to dorothy@wizard.oz.
Mail sent successfully to dorothy@wizard.oz.
```

The new character that we just created belongs to the default player class and is now ready to be used. If you want it to be able to create new rooms and other objects, you must make it a *child* of the generic builder class as discussed above. The command that is used to change the *parent* of objects in the MOO is called "@chparent."

@chparent Dorothy to $builder

Because our new character, Dorothy, is now a child of the generic builder, she has access to all the basic building commands in the MOO such as @dig, @create, @recycle, and many others. If builder status is all you want the new character to have, you can stop. In our example, however, we wanted to make Dorothy our new wizard character, and in order to do so we must first make her a programmer by using the @programmer command. This is a special command that will make new players into programmers directly without going via builder first, and if you want to give new players programmer status right away, this is the command you should use. What the @programmer command does is change the parent of the player to generic programmer and set a special property that tells the MOO that the player has extended programmer permissions. In addition, it sends a MOOmail to the player informing her of her new status and another mail to the MOO's New Programmer Log so the wizards can see who has been given programmer status. Thus, to make Dorothy a programmer we simply type:

```
@programmer Dorothy
```

In response to this command the MOO will produce the following output:

```
@Programmer Dorothy despite its lack of description?
  [Enter 'yes' or 'no']
yes
Dorothy (#100) is now a programmer. Its quota is
  currently 7.
Dorothy and the other wizards have been notified.
Dorothy is now a programmer.
*New-Prog-Log (#29) has just been sent new mail by
  Wizard (#2).
```

The final step in making Dorothy a wizard involves changing her parent to generic wizard and setting a special property called wizard to notify the MOO of her new extended powers. Since making new wizards is not something we do every day, there is no special command for this purpose. Instead we have to make use of the @chparent and @set commands we discussed earlier.

```
@chparent dorothy to $wiz
@set dorothy.wizard to 1
```

In addition to your new wizard character, you should also make a new programmer character for yourself that you can use on a day-to-day basis. The powerful commands and permission rights available to wizards may cause serious damage to your MOO if not used correctly and with proper caution, so it is always good practice not to use your wizard character unless you have to in order to create new characters and take care of other administrative duties. Until you get more experience with your MOO, you are therefore well adviced to err on the side of safety when it comes to using your wizard character.

Guest Characters and How to Make Them

Most MOOs will want to have guest characters so people can log on and see if they like the community before requesting a player character. Unlike registered characters, however, where administrators have access to the person's email address, guests are anonymous. This means that you have no way of knowing who they are, or holding them responsible for what they do in your MOO. For this reason some MOOs do not have guest accounts. As an administrator, however, you can easily find out what site the guest is logging in from, and based on this information you can effectively block access from that site. For more on this see the section on blacklisting below. Although the problem of obtrusive guests should not be underestimated, we believe that the advantages of having a guest system by far outweigh the potential problems. An important reason for this is that guest characters are reuseable and do not require passwords. This means that people who just want to log on to your MOO for one class or online meeting won't have to bother remembering yet another password, and you will be spared the work of having to create new characters for them—characters that may never be used again.

Since the guest characters are reusable, their names are composed of a freely chosen word or name followed by the word *guest*. You can pick any word you want, but we recommend that you choose words that go with the theme of your MOO. In ATHEMOO, for instance, which is devoted to theater and dance, guests are called Shakespeare's Guest, Dante's Guest and so on. The command for making guest characters is called @make-guest and should not be confused with the @make-character command that we discussed above. If, let's say, we want to create a guest character named Toto we should type:

```
@make-guest Toto
```

The MOO will now create a new guest account called Toto Guest. Note that the word *Guest* is appended to the guest name automatically when it is created and should not be typed in. Nor should you include any email address, because guests do not belong to any one person in particular and, furthermore, do not require any passwords. You may create as many guest characters as you want, but be sure to pick unique names for all of them.

Passwords and Security

Access to your MOO is regulated through a password security system. This system is your best protection against crackers and others who may want to break in and harm your MOO. Inside the MOO, users' passwords are stored in an encrypted format that is very hard to break, and no one, not even wizards, is able to read them. The real threat, therefore, is if someone manages to get hold of an uncrypted and readable password to your MOO somewhere and use it to log on. Although anyone can access your MOO via the guest system, guest characters do not represent any security risk because of heavy permission restrictions and the harmless set of commands they have available to them. Player and builder characters are also relatively harmless, whereas people with programmer status can, under certain circumstances, do quite a bit of damage either intentionally or accidentally. Needless to say, if anyone with ill intentions should obtain the password to one of the wizard characters in your MOO, there is no limit to the amount of damage they can cause. You should therefore be very careful to keep your passwords secure at all times, change them often, and never leave them in places where others may find them.

To change passwords we use the command @new-password. If you remember, the first time you logged on with the wizard character the MOO did not ask you for any password. This was because the makers of the LambdaCore wanted you to be able to log on without having to get a password from them first. Now that you are in and have created your own wizard character, however, you must set a password for the first wizard character. Failing to do so will leave the backdoor to your MOO wide open to anyone and represents a major security risk.

```
@new-password wizard is sOmEpa5SwORd
```

Do not choose passwords that are easy to guess, such as your birth date, the name of your cat, or the license registration on your car. The most secure passwords are those consisting of six characters or more. Because the MOO's password authentication system distinguishes between capital letters and small letters, you should also, as an extra safety precaution, choose a password that is a combination of capital letters, small letters, and numerals as illustrated in the example above.

A Few Other Important Wizard Commands

As a MOO administrator you have access to several commands for system administration and user management. There are too many to cover them all here, but in what follows we are going to take a look at some of the commands that you will be using most often. You can get a complete listing of wizard commands by typing 'help wiz-index' in your MOO.

`@quota`

Every time someone creates a new object (or digs a new room) in your MOO, the database gets bigger. Most objects are fairly small, so the increase does not affect the size of the database in any significant way. Over time, however, if everyone were free to create as many objects as they wanted, your database would soon become very big and potentially outgrow the RAM resources of the machine on which it is running. This phenomenon is called database bloat and can turn into a serious problem that every MOO administrator must be alert to.

To prevent uncontrolled database growth, the LambdaCore has a quota system that puts a limit on the number of objects builders and programmers can create. By default this number is set to seven. For many people who just want to create themselves a room with a couple of exits and a few other objects, this should be sufficient. For others, however, seven objects may be too few for what they want to build. As an administrator, you have the power to increase a quota at any time using the @quota command. In the example below we use this command to raise Dorothy's quota to twenty objects.

```
@quota Dorothy is 20
```

When people ask you for more quota, you should first take a look at what they have already created in order to determine whether to give them more. The @audit command shows you a listing of everything the person owns. In many cases you will find that they can free up quota by recycling objects that they own but do not use.

There is an alternative to the object-based quota, called byte-based quota (BBQ). Instead of looking at how many objects a person can create, the BBQ system measures the size of the objects a person owns. In large MOOs, where RAM resources are often stretched to the limit, a BBQ system gives you a much more exact estimate of the actual database size and helps you control growth better. The LambdaCore comes with both systems installed, so you can always switch to BBQ system later if database growth becomes a problem. The *High Wired* enCore comes with byte-based quota as the default setting.

`@shout`

The @shout command is used to broadcast a message to everyone currently connected to your MOO. Because many people see it as annoying to be interrupted by such broadcast messages, we recommend that it be used rarely. The command is available only to wizards.

`@@who, @net-who`

The commands @@who and @net-who show you a special listing of connected players, identifying the site from where they are logged on. This information can be very useful in case you need to deal with troublesome and unruly guests (see below).

`@blacklist, @redlist`

Most all MOO administrators must, at one point or another, deal with people who make trouble and harass others on the MOO. Sometimes these can be anonymous guests, sometimes registered players in your MOO. How you deal with these problems will vary from situation to situation, but usually conflicts can be avoided if you talk to the offender and ask him or her to stop whatever they are doing. Sometimes, however, the gentle approach will not work, and you will have to resort to other and more serious measures in order to protect your MOO users from harassment. If someone is creating problems in your MOO and, moreover, refuses to listen to your repeated warnings, you should first use the @@who command to determine what site the person is logged in from and then type the following command to disconnect the person from the MOO immediately.

```
;boot_player <object number>
```

Once you have done this, you should also make sure that he or she cannot log back on immediately to continue their unruly behavior. For this purpose you have the two commands @blacklist and @redlist available to you. The @blacklist command will disable player creation (if you have automatic player creation turned on) and all guest logins from the site or sites in question, while the @redlist command will effectively block all connections from the actual site or sites. To be absolutely certain that a site is effectively blocked you should blacklist or redlist both its domain name and IP number, for example,

```
@blacklist troublespot.edu
```

In most cases blacklisting a site for a few days or so should be sufficient to discourage the perpetrator from coming back. However, if this doesn't help and the person or persons keep coming back to harass people on your MOO, you should not be afraid to redlist the site they are coming from. Your first and foremost responsibility should be toward your users, and if this means blocking all access from particular sites, it is your duty to do so.

@newt

If you find that one of your registered users, a person with player, builder, or programmer status, is causing problems, you should first try to talk to the person and firmly request that he or she behave according to the code of conduct expected in your MOO. If this doesn't help, your next step should be to temporarily disable the person's MOO character with the @newt command.

```
@newt Wicked_Witch Refused to adhere to proper conduct
```

As in the example above, you can include a commentary explaining why the person was newted. The newting will be in effect until you use the @denewt command, and during this time, the person will not be able to use his or her character.

@toad

If, after a period of newting, the person still refuses to adhere to proper conduct and continues to pester other users and create intolerable problems for you, you should consider permanently deleting his or her character. Unlike all other objects, players cannot be deleted in the conventional manner using @recycle, so to permanently delete a character from your MOO you must use the special command @toad. In addition to deleting the character you may also wish to add the person's last connected site to the blacklist or redlist in order to invoke various kinds of blocking on that site and prevent him or her from coming back as a guest. Toading a person is a drastic thing to do and should be used only as a last resort.

```
@toad Wicked_Witch
```

To prevent the person from coming back as a guest the following two commands can also be used.

```
@toad! <character> is synonymous with @toad
  <character> blacklist
@toad!! <character> is synonymous with @toad
  <character> redlist
```

@dump-database

This command is used to save a copy of the database to disk. There should be no need to use it very often because the server makes automatic backups of your database every hour. This automatic backup procedure, also known as checkpointing, is extremely important. In the beginning of this chapter we said that the LambdaMOO server keeps the entire database in RAM at all times. This means that if, for some reason, the machine on which the MOO is running should go down, your most current database will be lost, and you will have to restart with the backup copy. Since the MOO checkpoints every hour, however, you shouldn't lose more than an hour's worth of work in the worst-case scenario.

The copy of the database made at checkpoint is stored in your MOO directory and named database_name.db.new. As an extra safety precaution you should make a backup of this file at least once a day by using the UNIX cp command. In the example below we are making a backup of the database file mymoo.db.new and naming it mymoo.backup. In the event that your regular database file should be corrupted, this extra backup will prevent you from losing your MOO.

```
cp mymoo.db.new mymoo.backup
```

@shutdown

The safest way to shut down the MOO and stop the server is the @shutdown command. When you use this command anyone who happens to be online is given a two-minute warning that the MOO is going down so they have time to quit whatever they are doing and log off. When the two-minute warning period is up, the MOO will disconnect all users and copy the database to disk before sending a quit signal the server program. If you need to abort the shutdown process for any reason, use the @abort-shutdown command.

The MOO Newspaper and Other Mailing Lists

The MOO has a convenient and useful news and mailing list system that you should learn to use and take advantage of. The MOO newspaper, for example, is a convenient place to make announcements that you want everyone to see. To use it, simply send a MOOmail to it and then add the new article to the current edition, as illustrated below. If you are not familiar with the MOOmail system, please read about it in chapter 3. All mailing lists, including the newspaper, are referred to with an asterisk (*) prepended to their name (e.g., *news). So in order to send a new article to the newspaper we type

```
@send *news
```

This invokes the MOOmail editor so we can compose our message, while the following command will add it to the current edition of the newpaper as article number 2. You will have to actually mail the message to *news by typing 'send', before you can use the @addnews command.

```
@addnews 2 to *news
Current newspaper set.
  1:   Feb 11 09:20   Wizard (#2)     Welcome to
                                      LambdaCore
  2:   May 11 22:34   Wizard (#2)     Greetings from Jan
                                      and Mark
----+
There's a new edition of the newspaper. Type 'news new'
  to see the new article(s).
```

The *High Wired* enCore comes with a different news system, and we refer you to its web site for more information on its use. In addition to the newspaper, which is reserved for the wizards' use only, you can also set up a number of other mailing lists or discussion groups that anyone can use. In many MOOs, for example, there is a list called *general that players and wizards use for announcements, general discussions, and more. In the following example we show you how to set up mailing lists by making a new list called *general; and remember, you should type in only what we have boldfaced. Note that the @add-news command is only used when sending new articles to *news, not any other mailing list. See chapter 3 for more information on how to subscribe to and post to MOO mailing lists.

```
@create $mail_recipient named General
You now have General with object number #103 and parent
  Generic Mail Recipient (#45).
@describe General as This is a mailing list for general
  announcements and informal discussions about what
  goes on here in our MOO. The list is open to anyone who
  wishes to subscribe.
Description set.
@set General.readers to 1
Property #103.readers set to 1.
@move General to $mail_agent
You teleport list.
@subscribe *General
*General (#103) has 0 messages. Notification of new
  messages will be printed when you connect.
```

In this example we created a public mailing list that anyone can read and post to. Sometimes, however, you may want to create mailing lists that only wizards can read. For example, on most MOOs there is a list called *wizards that the administrators use to discuss various things among themselves. If you want to set up such a list in your MOO, repeat the example above, naming your new list something like *Wizards, and skip the command '@set General.readers to 1'.

Designing the Public Space

Now that you have your MOO up and running, the core has been customized, and you have made your first few characters, you can get to the fun part of actually building new rooms and objects in your MOO. At this point you probably have a pretty good idea about theme, design, and layout of your MOO, and as you get to work on this there are a few design issues that you should bear in mind.

In every MOO there must be a public, common space from where others can start to build their parts of the MOO. This common space is also typically the first people see when they log on, and it is therefore important that it be as informative and comprehensible as possible. For this reason, many MOOs like CollegeTown, Lingua, Meridian, and others have designed visual ASCII graphic representations of the textual landscape (fig. 9).

Regardless of whether you choose to use graphics or not, you should try to make the public spaces in your MOO easy to navigate. For a new and unexperienced MOOer there is nothing more frustrating than constantly getting lost and not being able to find one's way around. A good rule of thumb, therefore, is always to give exits going back toward the starting point the name *out*. People who get lost can then always rely on the fact that typing 'out' will take them one step back to where they started from.

As in the example from Lingua, you should also make help texts and other information easily accessible and close to the starting point where people log on. Since the starting point is often crowded with people logging on or off or just hanging out, you should make a separate room where people may browse your help texts without being disturbed.

The commands that you will be using most frequently to build your MOO are @dig, @create, @describe, and @recycle. These commands are explained in chapter 3, so we will not repeat them here. Also remember to use the online help system if you need information on commands that are not explained in this book. Typing 'help <command name>' will bring up the help text on that specific command if there is any.

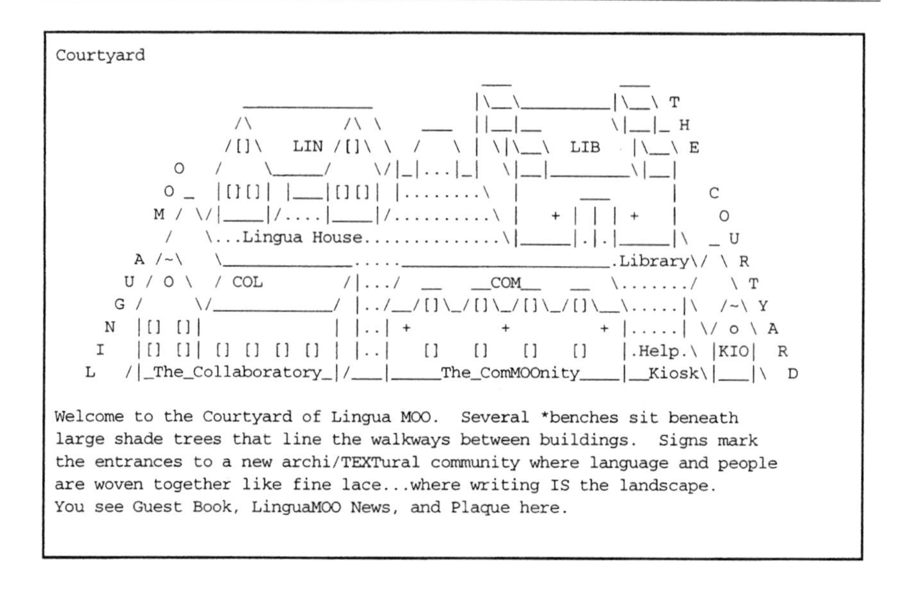

Courtyard
LIN
LIB
...Lingua House...
.Library
COL
COM
.Help.
KIO
/|_The_Collaboratory_|
The_ComMOOnity
Kiosk
Welcome to the Courtyard of Lingua MOO. Several *benches sit beneath
large shade trees that line the walkways between buildings. Signs mark
the entrances to a new archi/TEXTural community where language and people
are woven together like fine lace...where writing IS the landscape.
You see Guest Book, LinguaMOO News, and Plaque here.

Fig. 9. The Courtyard of Lingua MOO

Porting Objects from Other MOOs

Since most MOOs are based on the LambdaCore, you can transfer objects between them relatively easily. Copying objects between MOOs is generally referred to as *porting* and is something that you will almost certainly want to do at one point or another.

The database that we provide for use with this book comes fully equipped with many educational tools and features, something that should eliminate the need for immediate extensions of the MOO. If you are using the generic LambdaCore, however, these and other tools will have to be imported into your MOO. Depending on what client program you are using to connect to your MOO, there are several ways to port objects from one MOO to another. There are, however, some general steps that must be followed, and in the following example we illustrate these by showing you an easy way to port using Telnet in a Macintosh or Windows environment.

If, let's say, you want to port a recording device from SomeMOO (A) to your MOO (B) you should first open two simultaneous Telnet connections, one to MOO A, and one to MOO B. The next step is to determine what parent the object you want to port from MOO A has. The command for determining this is called @parent, so you select the window to MOO A and type

```
@parent recorder
```

```
Recorder(#517) generic thing(#5) Root Class(#1)
```

The @parent command tells us that the recorder is a child of the generic thing. Once you have found this information, you should switch back to your MOO and create a new object with the same parent as the one you want to port from MOO A.

```
@create $thing named Recorder
You now have Recorder with object number #104 and parent
  generic thing (#5).
```

Now, make a note of the number of the new object (#104) that the MOO just created for you, and switch back to MOO A again. We are now ready to start porting. The command for this is called @dump, and here is how we use it.

```
@dump Recorder with id=#104 create
```

The first argument tells the MOO which object we want to port. In this case it is the recorder (which could also have been referred to by its object number). The next arguments tell the MOO that we want the new object that we are eventually going to create to have object number 104, the number we got when we created the object that is going to become our new recorder. When we press return/enter to execute the command, the MOO spills out a long listing of properties and verbs that are defined on the recorder we are porting. Now, select the whole listing and copy it by using the 'copy' command of your Telnet program, then switch back to your MOO.

The process of re-creating the recorder object in your MOO is simple. All you have to do is select 'paste' from your Telnet menu to paste the listing into your MOO window. In response the MOO will tell you that it is adding properties and programming verbs. When you see a line saying "You say, '***finished***'," the porting process is done and your new recorder object is ready to be used.

The last thing you need to do is to recycle an extra object that was created in the porting process. If you look at your inventory (type 'i') you will see that you are carrying two objects with identical, or almost identical names. The object that you want to keep is the one with the object number that we made a note of earlier (#104); the other one is redundant and should be recycled. Be careful, however, not to recycle the wrong object, or you will have to repeat the whole process from the beginning.

One very important thing to remember when you port objects from other MOOs is always to ask for permission. Some objects take a long time and a lot of hard work to make, so out of respect for the programmer who made them you should always ask before you port them to your own MOO.

Choosing the Wizard Team

As the head administrator or archwizard you will at some point have to recruit a team of people to help you with the daily operation of your MOO. A few MOOs, such as LambdaMOO, MediaMOO, CollegeTown, and others, have experimented with various models of online democracy and user governance. Although this may be an interesting and fruitful experiment, especially in socially oriented MOOs, we feel that the best model for an educational MOO is one in which there is a small group of specially appointed and responsible administrators. It is impossible to give you any prescription on how to compose the best possible wizard team, but based on our years of experience in MOO administration we can offer some advice and recommendations.

As a general rule you should compose your wizard team so that it is as small and efficient as possible. Also make sure that each person has a clearly defined duty. As the head administrator you must take the ultimate reponsibility for supervising the whole operation of the MOO. In addition you should have one or two people whose main responsibility is to look after the technical operation of the MOO server and database. One person could handle all character creation and quota policies; perhaps another person should deal with questions and problems that the players have. Thus, for a large or medium-sized MOO, a wizard team of four to six people should constitute an ideal size, while for a small start-up MOO a smaller group of two to three people should suffice.

Being a wizard carries not only absolute powers in the MOO world, but also a great responsibility. As the owner of the MOO you must therefore make sure that the people you choose for your wizard team can be trusted and that they know how much work is involved. Ideally, you should only choose people that you know well and that you know to be mature and responsible. Do not ever choose a person to become a wizard in your MOO just because of his or her technical knowledge or skills. Once you have wizzed someone, you have given that person access to every bit of information in the MOO, including commands that might be used to destroy it. So before you make the decision to make someone a wizard you should trust and know that person completely. By being careful and selective about whom you choose as wizards you will avoid many problems and internal conflicts that will certainly make the job easier for you and also give your educational MOO a good and professional reputation.

The Ethics of Wizardhood

As we have emphasized several times in this chapter, being a MOO wizard carries a great deal of responsibility that needs to be taken very seriously. As archwizard you are the mayor, chief of police, and judge of your little community, and needless to say this can pose quite a few dilemmas with regard to ethics and the ideal of democratic governance.

In spite of this there is one golden rule that every MOO wizard must abide by, and that is not to abuse one's wizardly powers. This means, among other things, that you should respect your users' privacy and not eavesdrop on conversations or read their MOOmail. You must not move or recycle any of your users' objects without their explicit permission. Do not ever move players around without their express consent. Always answer questions from your users in a polite manner, and don't be rude; remember that even the most experienced MOO wizard was once a newbie.

We have in this chapter touched on some of the most important things that a new MOO administrator needs to know. Although there is much left to learn, you have taken the first steps on the long and winding road to becoming an archwizard.

Internet Resources for MOO Administrators

The LambdaMOO Server and Database Archive. On this FTP site you can always find the latest version of the LambdaMOO server and core database as well as other MOO related information and utilities. URL: ftp://ftp.lambda.moo.mud.org

The *High Wired* enCore. This educational MOO core database is a constructivist virtual reality environment designed specifically for educational use. It is based on the February 2, 1997, version of the original LambdaCore Database and supplemented with a number of useful educational tools. The *High Wired* enCore also comes with a built-in web-interface, support for client interfaces such as the Java-based Surf and Turf interface and MacMOOSE, in addition to numerous other features and enhancements. On the *High Wired* enCore web site you will also find references to many useful online resources for MOO administrators. The *High Wired* enCore is provided free of charge. URL: http://lingua.utdallas.edu/hw/encore.html

LPMOO Home Page. LPMOO is a disk-based MOO server developed by Robert Leslie that offers an alternative to the LambdaMOO server. For more information on LPMOO visit the LPMOO web site. URL: http://www.mars.org/home/rob/lpmoo.html

The File Utilities Package (FUP) FTP Site. FUP, written by Gustavo Glusman and Jaime Prilusky, is a way for the MOO to actually use the file system that UNIX provides. It can be used for many things including taking mail and news groups out of memory and storing them on disk so that the MOO itself does not take up as much RAM. More information can be found at the FUP archive site. URL: http://bioinformatics.weizmann.ac.il/software/MOOsupport/FUP/

The Official MOO-Cows FAQ. Web site with answers to frequently asked questions about MOO and MOO administration. Maintained by Ken Fox. URL: http://www.ccs.neu.edu/home/fox/moo/moofaq.html

The LambdaMOO Programmer's Manual. Pavel Curtis's specification of, and manual to, the internal MOO programming language. It is quite comprehensive and may be hard to understand if you are not familiar with computer programming, but it is a document that every MOO administrator should know about. This manual is also available in text and postscript formats on the LambdaMOO FTP site listed above. URL: ftp://ftp.lambda.moo.mud.org/pub/MOO/ProgrammersManual.html

The MOO-Cows Mailing List. This is an email discussion list for MOO administrators to discuss technical questions related to the LambdaMOO server and database. The list is primarily for MOO administrators, but others are also welcome to subscribe. To subscribe send an email to: Majordomo@the-b.org with the words subscribe Moo-Cows in the body of your email.

The MOO-ed Mailing List. This mailing list is for discussion of the programming, maintenance, and pedagogy of educational MOOs. The list is open to anyone who is interested in discussing these issues. To subscribe send an email to: Majordomo@ucet.ufl.edu with the following command in the body of your email message: "subscribe moo-ed <your email address>"

Quick Reference to UNIX Commands Used in This Chapter

cd	Changes to the given directory or path.
dir	Comprehensive listing of directory contents.
gzip	Invokes the compression and decompression program.
ls	Abbreviated listing of directory contents.
make	Invokes the UNIX C-compiler.

man	Brings up the UNIX manual pages.
mkdir	Makes a new directory.
mv	Renames a file.
pwd	Lists the path of your present working directory.
rm	Deletes a file.
tar	Invokes the UNIX archiving and unarchiving program.
vi	Invokes the UNIX text editor vi.

6

Day-to-Day MOO Administration and How to Survive It

Shawn P. Wilbur

For its administrators, every MOO is an educational MOO. This is inescapable, and we forget it at our peril—or at the peril of our MOO-based endeavors. The same factors that make a MOO so inviting as an environment for collaboration and education, most importantly its flexibility and openness, make adminstrative prescriptions very difficult. For better or worse, nobody can provide you with a foolproof blueprint for running your MOO. You can expect to learn to manage your site on the job. That is not to say that there is not a great deal that you can learn about MOO administration from those who have already gone through the process of developing an administrative regimen. It does, however, shape the sort of information that can be passed along. In what follows, the emphasis will be on the insights and skills required to make your on-the-job training less painful, for you and others at your site.

This chapter is designed to help administrators learn how to think about MOO, and the various tasks involved in maintaining a site, so that they can develop and maintain routines appropriate to the specific needs of their own projects. It begins with an administrator's-eye view of MOO, an overview focusing on aspects of the software and the interactions it fosters that shape the general nature of administration. The rest of the chapter is dedicated to an analysis of several broad categories of tasks—planning, creation, and maintenance—in both their technical and social aspects. There will be inevitable overlap with other chapters, and perhaps also inevitable disagreement. Feel free, even obliged, to take the strongest statements with a grain of salt. The view presented here is finally only one among many ways that MOO administration can be approached. However, it is one based on considerable personal experience of unexpected growth and conflict in MOO-based environments and is designed to be a relatively safe program. If your ambitions are small, and the site you envision simple, then you may be less pressed by the peculiarities of online interaction. But no site that is not both extremely simple and rigidly controlled can be expected to develop without throwing its creators and administrators the occasional curve. And sometimes the surprises can be extraordinary, or catastrophic.

Moo Theory for Administrators

Much of the appeal of MOO lies in its relative simplicity. A few simple commands suffice to enable participants to move about, communicate with others, and manipulate, create, and customize virtual objects. The internal scripting language allows even unexperienced programmers to begin to shape the virtual spaces in subtle ways fairly rapidly. And while the text-based form of virtual reality has its limits, it is hard to find a more powerful language for shaping a world than the languages we use every day—and these are languages with which even our novice users can be expected to have a considerable proficiency. We need only scan the shelves of a bookstore or library to gauge the flexibility of ordinary language to shape virtual worlds. Much of MOO's power remains untapped, still awaiting the writer-programmers who can unlock more of its potential. On the other hand, the range of specially tailored environments based on MOO has become extensive. The educational core designed to support this book includes some of the more elegant adaptations of MOO, objects that have become popular on various sites—thanks in part to the relative uniformity of MOO-based sites, which allows objects to be ported fairly easily from one to another.

We might think of MOO as a very open system. Set this system in motion, however, and the picture can change dramatically. When a MOO is in active use, all the simplicity, flexibility, and openness conspire to create a dizzyingly complex affair, which will undoubtedly be prone to conflicts and crashes, both local and global, minor and catastrophic. The key to survival in MOO is the ability to see, and take comfort from, the simplicity that still underlies all the daily chaos. (Take this as a mantra: Moo is a system of simple systems.) This ability is important to administrators and other participants alike, but you may find that grace under virtual pressure is a relatively scarce quality. There are good reasons for that scarcity, and they have everything to do with the way that MOO structures experience. Although a serious look at the psychology of being in text-based virtual reality is beyond the scope of this chapter and would in any event still be premature, it may be possible to pose a few basic problems of MOO identity and society, specifically as they relate to administrative concerns, as a way of determining what needs to be addressed in the philosophy of MOO administration that every "wizard" (or "janitor" or "staff member") will need to compose for themselves and their site.

Managing the Virtual, Teaching the Virtual

In MOO, we *are* the virtual as much as we are *in* the virtual. That is, we encounter one another only to the extent that we successfully translate into text ourselves and others and a world that we presumably share with them. A great deal is lost in this translation, but there remains the usual noise added to the flow of information on which our presence to one another depends. We often forget this, even in text-based environments. No matter what the scoffers say, MOO can be tremendously immersive.

Text is a peculiar medium, poorly understood and undervalued in the age of electronics. Perhaps the influx of educators and students into these strangely rich textual environments will offer opportunities for understanding and teaching a new grammar and poetics of electronic text. Or perhaps, as we confront one another already transformed by the medium, we lose a critical distance we presume between text and identity. Perhaps only time will tell.

Consider the processes by which we assume an identity in MOO. There are complex forms of misrecognition involved. We see the text on the screen but recognize a much broader range of experiences and sensations. We "hear" speech or are distracted by noise, and we "see" text translated into images of greater or lesser clarity. We even experience various sorts of tactile contact, though this sort of benign synesthesia presses the limits of text-based virtual reality. Various sorts of intense communicative activities seem to conjure bodies up out of text, and it is possible even to create sensations of movement when our physical bodies are stationary, seated in front of the computer. We sense that there is something like an exchange between the player object and the participant, by which the object is infused with life by its owner, while that owner must acquiesce to the rather rigid rules of the server.

Presence in MOO is essentially impossible if the participant is not also an active reader and interpreter, and interpretation in this case means, among other things, the translation of text into being. Translation in this sense is creation, a filling in of details and a more or less willed reorganizing of our expectations of what it means to be somewhere with other people or objects. Reading is writing, or creating, and every participant plays a role in shaping the environment, whether or not they ever create an object or program a single verb. In some sense, every participant shapes the whole environment—not without limits, but perhaps in ways that differ considerably from other participants.

Administrators must consider the possibility that their sites exist as a multiplicity of overlapping worlds, each highly personalized, which cushion the various participants from the sparceness of the MOO environment and the demands of the "real world." Immersion in MOO seems to depend in part on a kind of functional solipsism. This view may provide some insight into the volatile social dynamics of MOO societies. These personal worlds rub against one another with little friction, for the most part, but when conflict intensifies, it is as if everyone's bubble world threatens to burst at once. It is no surprise that in the heat of social strife, MOO users often seem to be speaking different languages, since they are in some real sense inhabiting different worlds.

MOO is a technology of the *virtual* precisely because of its flexibility, because its reality can be simultaneously actualized in so many relatively compatible ways. Without this dynamic, it would not be nearly as powerful an environment for collaboration, but because of this dynamic, collaboration must come from a conscious transformation of separate cocreation into joint work that shares common values, goals, and language. It is finally necessary to *manage* the fragmentation and uncertainty that come with virtuality, rather than attempt to eradicate them. In part, however, this means that uncertainty must be recognized and made a part of the common model. So administrators will need to make education about the possibility and constraints of the environment a priority, in addition to education in whatever field or fields the site has been designed to support. This is a difficult but exciting task, a perfect challenge for educators in the humanities, and a spur to developing exciting forms of literacy.

Some readers of a more practical bent may wonder if too much is being made of the issues of virtuality and online identity. Certainly, there is a common-enough sense that technologies are "just tools," and that to mix concerns about the technological environment with concerns about the content of our educational endeavors might well be to err by mixing unlike things. The voice of experience says this: teaching in MOO is likely to be as radical a change in circumstances as teaching in a bubble on the moon, or outside in a blizzard. Most teachers I know hesitate to take their classes outside on a sunny day, for fear of shattering the fragile learning atmosphere we work to build in the classroom. To enter an online setting without at least an equal dose of forethought and trepidation is probably just plain foolhardy. On the other hand, it is easy to take the classroom for granted and to rely on classroom habit to carry educational motivation, if we never take the chance on a session on the grass, or in cyberspace.

Memory, Sprawl, and Focus

Fortunately, at least for those who find negotiating the niceties of online identity-maintenance tiresome, the other major economies of MOO answer to a logic that is much less arcane, if no less inclined to crisis. Unless you curtail growth of the MOO in some way—and all such limitations risk curtailing learning and creativity—your site will swell until it uses all available memory, causing increased lag and then crashes. Of course, it may even sprawl more rapidly according to individual players' whims, and without any collaborative plan, until large portions of it are unnavigable. Unless you are gifted with extremely generous system providers, memory limits may be fixed constraints. The first type of growth problem will probably have to be dealt with by occasional, or ongoing, database-reduction campaigns, which are covered in a later section of this chapter. The possibility of sprawl and loss of focus is primarily a problem of policy and planning. If you plan to allow participants to add to the terrain of your site, or swell the database with other sorts of objects, then be sure to establish guidelines or even administrative bodies to keep that growth within limits that will not tax your site too sorely. Modifications to the @dig or @create code, limiting the circumstances of new building, or the creation of quota or architecture review boards, are possible solutions to choosing between free creation and no creation. Making certain that the site's theme and goals are clear and well publicized is another approach to the problem.

The difficulty is maintaining clarity in a changing environment. Imagine the development of MOO as a movement with three moments—planning, creation, and maintenance—in multiple, concurrent cycles. Planning, in this model, includes not only all the necessary preplanning, prior to creating and opening a site, but also frequent reassessment and replanning, to fit your administrative master plan to the current realities. Similarly, creation is a constant element of the cycle, as new objects and classes are created, and as new participants take a turn in the complex dance of identity creation. Maintenance under these circumstances will not be a matter of maintaining a given set of conditions in a static state. Instead, it will be a matter of maintaining continuity, and minimizing disruption, in the midst of change. Again, it is largely a matter of managing uncertainty.

Why Development?

Of course, given the difficulties that this active development seems to create, you might be tempted to clamp down on growth and hold things static. If you have planned well in advance and know what you want to do with your site, why shouldn't you attempt to hold things static? The main reason is that it does not seem to work, for a variety of reasons that are still poorly understood, but among which three stand out.

First, you will have to be concerned about development because it is almost certain that your users will be concerned about it and may even be resistant to investing time and energy in a text-only environment, particularly if your site is not producing a fairly steady stream of novel objects, spaces, and experiences. Computer users are conditioned by a market that turns over technologies more and more rapidly. Students and educators will probably acknowledge that the thirst for new "cool tools" has not been absent from academe. For better or worse, most MOO-based sites are in some sense subject to that market's logic, competing at the very least for the attention and active involvement of participants. So these are demands that administrators can hardly afford to simply ignore.

Fortunately, a second reason for being concerned with development is that new code, new activities, new interfaces to other resources, and new participants all provide opportunities for the kind of (most frequently minor) course corrections that keep a MOO functioning relatively smoothly. Novelty can shake up the site's business as usual. Public discourse about development not only helps to clarify problems but can clarify individual perceptions in way that may help create consensus and reduce conflict. As long as the new additions to your database are not simply more of the same—quantitative increase without the addition of new qualities—then most sorts of development can be turned to your site's favor.

A third major reason that administrators should concern themselves with development is that it represents an opportunity to hone their skills, and a compelling reason to ask, and ask again, difficult but important questions about the purpose, status, health, and direction of their site. And, lest we forget, it can be fun, exciting work. Administrators, like other participants, will probably respond positively to gradual changes in their routine, to the introduction of novelty and to the feeling that their endeavors are "going somewhere."

Just recall that few notions are more ambiguous than development. We have already acknowledged that one sort of development involves the constant production of novelty, with little distinction made between useful novelties and others. This sort of development can institute a cycle likely to turn vicious for MOO administrators. Indiscriminate growth is a luxury few sites can afford, and for which they may pay in increased lag and vulnerability to crashes. So it is probably best to focus less on novelty, which will occur unless you strangle creativity entirely, and try to focus on planned development with frequent reassessments of the results and corrections when they seem to be required. And one might well ask whether the thirst for absolute novelty is not, at least in many instances, an attempt to compensate for a lost sense of purpose and direction. Perhaps the paradoxical simplicity of MOO makes it one of the spaces in modern society where focus and purpose can be constructed creatively, with relative freedom, and maintained, provided there are those willing to work at the task. And perhaps its unreality, its at least apparent distance from the "real world," gives it certain advantages over the classroom as an educational setting, as educational institutions become increasingly integrated into corporate culture and respond to its values and priorities.

Getting on Task

By now, you should have the beginnings of a theoretical or philosophical image of MOO. But how, you might ask, does theory translate into routine practice? What are the specific tasks that constitute MOO administration? Or, to ask a better question, how does an administrator recognize which tasks have to be done at the given moment?

The first duty of an administrator is to maintain an active presence on the site. Without spending a considerable amount of time there, it will be nearly impossible to identify any but the grossest and most catastrophic problems, and it will be almost entirely impossible to prevent them. The general social scheme of your site will determine the sorts of interactions you engage in, but it is important that you be known and considered approachable. In a rough period on a MOO I help maintain, a parent of one of the widely used generics was inadvertently destroyed, but because communication between participants was disrupted by internal strife the event was not reported for several weeks, although nearly one hundred participants were affected. By contrast, in happier times on the same site, I was receiving enough friendly, unsolicited reports of unusual events that I was able to track a major security break before any real damage was done. Other kinds of active involvement, such as programming, involvement in activities, or mailing-list discussions, helps establish rapport with other participants, and with the system itself. But when problems do surface, the tasks they give rise to are likely to fall into a few basic categories: repair work, database trimming, and diplomatic work.

Fixing What's Broken

On the technical side, fixing what's broken is relatively simple, although it can be time consuming. There are a fairly limited number of ways in which important code is likely to be "broken," and when this occurs, the MOO server usually returns fairly complete information about what went wrong. These *tracebacks* identify the object, verb, line number, and type of error involved, together with a complete history of the various verb calls between the typed command and the error. Correcting the problem is a matter of tracing the flow of data back from the final verb call, which appears in the top line of the traceback, looking for the point at which the system is likely to have become corrupted. The ability to read and use tracebacks is a skill no MOO administrator should be without. Fortunately, it is a skill that is fairly easy to master. Naturally, the greater your familiarity with the workings of MOO code, the more rapidly you will be able to isolate and fix problems. But new administrators may find that the focused work required to read tracebacks is an efficient way to gain greater insight into those workings.

Beyond the information found in tracebacks, a great deal of information is available in the standard help databases and in comments on many of the more important verbs. As is nearly always the case, the documentation provided is far from perfect, and learning to find the bit of information you need when you need it is mostly a matter of practice. But new administrators might take the documentation provided as a mark to match or exceed as they begin to customize their own sites. The greatest challenges come when some new, undocumented, or poorly documented object begins to give tracebacks. And if you are making any major changes to key core objects, you can expect that there will be bugs to work out. Some sites have standards for verb documentation, object help messages, and help database entries, which must be met before new code or objects are put into wide use. Administrators should consider imposing such standards on themselves, particularly for critical projects. You may have perfect confidence in your colleagues' abilities to write robust code, but you may also regret such simple trust in the event of a problem. Programming styles differ as much as writing styles, and making sense of work written in an unfamiliar style can be maddening. Much time can be lost trying to read sound, but undocumented, code. And time is one of the resources that is difficult to renew in online environments.

When we begin to talk about the results of broken code in terms of lost time, we begin to see that technical issues are not readily separable from social issues. In MOO, social systems break at least as often as hardware or software, though these failures are sometimes less spectacular—and there is no server mechanism to provide us with some clear information about their sources. We have to fall back on our general sense of what it means to be in MOO and then extend that uncertain knowledge with the insights of sociology, management and communications theory, and psychology. Obviously, there is no space to cover all that ground here. Administrators in virtual settings will have to do their homework just like administrators and managers in other fields. Experienced teachers will at least have experience of classroom dynamics to draw upon.

On Being the Right Size

Lack of additional memory will eventually put an end to uncontrolled growth, but nobody will enjoy the way it happens. Increased lag and an increased tendency to crash are among the symptoms of a bloated MOO. The first will make nearly all activity extremely difficult, battering the sense of presence and eating up time, and the second will often result in lost work for programmers and lost time for all participants, not to mention extra work for administrators. Memory use will increase in two ways, through the creation of new objects and code in the database, and through the increase in space taken by the server's processes, often itself the result of database bloat. The second sort of bloat can be taken care of by shutting down and restarting the server. But database management is a constant priority, involving a series of difficult choices.

How seriously you take the task of trimming the database, and what standards you set for the process, will probably depend on how close to the limits of your available memory bloat has brought you, and how rapidly. When the time comes to slim things down, the simplest place to begin is with mail. Some sites have systems that purge personal mail after a given interval. Some encourage participants to have mail forwarded to their email accounts. Reducing the amount of personal mail in the database can greatly reduce its size but may inconvenience participants. You will have to weigh the costs and benefits of systematic mail purges, perhaps after determining the success of voluntary database-slimming campaigns. The advantage of the latter, if you can make them work, is that it involves participants with problems that concern everyone on the site. Sometimes, a willingness not to solve problems for everyone can pay off in greater popular satisfaction with the solutions that are eventually chosen.

Public mailing lists also afford opportunities for simple, large-scale database reductions. It is fairly common to archive older messages outside the MOO, where they are available via the World Wide Web (WWW), gopher or file transfer protocol (FTP). The primary disadvantage of this practice is that it removes important historical documents from sites that may already be susceptible to the rapid effacing of their past. If you choose to offload messages, be certain to clearly indicate where the archive is located. Mailing lists all have descriptions, and this information might be included in each list's description. Sites with active network connections and gopher or WWW slates could easily assemble links to their archive among their default links.

Small reductions in database size may also be accomplished by consolidating redundant objects or recycling obsolete ones. Feature objects seem particularly prone to proliferation without much thought about whether their features are provided elsewhere, and voluntary consolidation drives may meet with some success. Consolidation may also be an option for generic objects of various sources. Of course, most of these problems can be dealt with before the fact, by providing a central space for displaying generics and feature objects. In any event, the space created in this way is likely to be minimal, and the best reason for consolidation is probably to reduce confusion. This, of course, may be reason enough.

Unfortunately, one of the greatest causes of database bloat is the creation of new accounts. You can expect your participant base to grow unless steps are taken to curtail that growth. And eventually you will probably be forced to make space for new users by removing some old, hopefully inactive, ones. Administrators need to make decisions about how critical their memory needs are and develop appropriate guidelines for determining what participants and related objects will be subject to "reaping." Few sites reap participants that have been idle for less than three months, and many attempt to notify users prior to deleting their accounts. A little common sense goes a long way when developing and applying these guidelines. Remember that student accounts may be inactive during the summer months. Educational sites may want to be particularly cautious in late summer and early fall. Unless your site is huge, automating reaping tasks is probably not a good idea. If someone is not personally reviewing each account as it comes up for termination, you may find that popular generics, or even other administrators' back-burner projects, have suddenly disappeared. Although a standard period of idleness is a good way to begin selecting accounts for elimination, a set of standard exceptions is a good idea as well. (I am personally slow to reap very early or "old" participants, those owning popular generic objects, and those who are administrators at other sites. Each of these categories seems to call for an extra degree of courtesy and tact.)

Fighting Fires and Managing in the Middle

Finally, after the server and database have been whipped into shape, administrators will find themselves inevitably left with problems of communication, governance, and discipline. Recalling what we have said about MOO identities, it should be clear that strict, static, authoritarian, or confrontational approaches to human-resource management or conflict resolution will have their limits. For similar reasons, simple forms of democratic or representative governance also face great difficulty. Successful administrators are likely to find themselves in a shepherding and conciliating role. A little gentle guidance can go a long way in MOO, and much of the daily business of MOO administration will be a matter of small fixes, slight course corrections, and gentle, constructive suggestions. There are, however, limits to the tolerance that administrators can show. Even the most anarchic of sites still has to take care not to run out of memory, and it is a rare MOO that does not have to answer to university or computer system administrators for the resources that it uses. It is rare that the power to make decisions about hardware allocation or staffing for external maintenance will be in the hands of those who run the MOO. And it is unfortunately also rare to find computer support personnel or school administrations as eager to defend the sort of free expression that is often taken for granted in online settings. MOO administrators are required to mediate between the "world" internal to the site and the institutions that surround it and supply it with resources, and this can require accomplished persuasive technique. It can feel like tightrope walking when it is time to request new memory for the machine housing your MOO, and someone on the appropriate board asks, "So what is this, er, MOO? I've heard about these chat sites, and . . . " But MOO *is* a powerful technology, and if your site is working, you will be able to prove it. Careful cooperation and delegation within administrative teams and carefully cultivated rapport with both MOO users and institutional administrators are among the skills needed to make that mediating position feel like "where the action is," rather than "caught in the crossfire."

Go Your Own Way

There are a thousand and one tasks, most of them minor, that you will learn to deal with as an active MOO administrator. Obviously, this chapter has done little more than scratch the surface of the mountain of details that keep a MOO up and running. But the details will come, and they will come more easily if you are philosophically prepared for the vagaries of online interaction. An awareness of the complex nature of online identity, a clear theme and development plan, and a set of clear criteria for database reduction will smooth the road for you considerably. Beyond that, MOOs rise and fall on the managerial and educational talent of their administrators and participants, and on the willingness of all to work toward building a sustainable whole.

III

Educational and Professional Use of MOOS

. . . Or does it make things harder to imagine?
Seeing inside the circus, they say, robs it
of romance. So you leap from ring to ring, fixed
on finding the New Circus. The New Circus

is a circus of infinite rings, led by
invisible wizards with magical fingers.
The animals of The New Circus are made
of dreams and current; they dance on electrons.

And the tightrope walker of The New Circus
amazes the crowd, which surrounds him rather
than kneel below, by stepping off his high wire
into the next dimension without falling.

Some will say The New Circus is not a real circus.

—Brian Clements

7

Help! There's a MOO in This Class!

Cynthia Haynes

In teaching there are unplanned, felicitous moments when something extraordinary happens in our classrooms. Sometimes those moments occur outside the walls of our classroom: a sunny day prompts us to move the class outside, or we take a trip to the library, or a seminar is held in someone's home where food and drink accompany the informal discussions. Often as not we wonder how to replicate that experience, how to capture it, how to reproduce the magic. Somehow, between the 1960s revolution and the age of digitized books, we found ourselves trapped in the conventional classroom, haunted by stifled learning, oppressive seating arrangements, time and space—boundaries we long to transgress. And then there was MOO.

In previous chapters, you have become familiarized with the general context for educational MOOs within virtual communities, and how to create and administer one. We are ready now to explore various ways that MOOs are being used in teaching. Our focus shifts, then, to concentrate on applications of MOOs in educational settings. In the process of shifting, we will shuttle between theory and practice, acknowledging that the one informs the other. We will assume the perspective of teachers new to using a MOO in their pedagogy, anticipating questions for consideration that will undoubtedly arise during the course of each MOO classroom session. Thus, the chapter is set up to be useful over a period of time, beginning with preparation for first-time MOO use to more advanced applications that are not bound by class time or space. Along the way you will find detailed activities based on successful experiences of teachers who have used MOOs in teaching, as well as suggestions for evaluation of MOO activity, and for handling inappropriate conduct, questions often foremost in our minds as we ponder the introduction of MOO activities in our pedagogy.

Preparing Yourself and Your Students

Once you have decided to take your class to a MOO and have acquainted yourself with the MOO, it is crucial to make sure that your institution and/or your department allows access to MOOs. If not, this book would make an excellent argument for understanding and valuing educational MOOs. You may also want to email our other authors for outside support in changing the policies at your institution. Teachers need to be as familiar as possible with MOOs in general and with a specific MOO before bringing their classes online. As a MOO administrator, I can attest that this point needs to be made. Learning the basic commands and learning how to navigate your way through the MOO is a minimum requirement before holding classes. This is not necessarily a time-consuming process. Most educational MOOs have excellent tutorials online and help texts to orient teachers to the basics. Of course, the more you explore, the better prepared you will be to help your students during their first MOO sessions.[1]

As a first step, it is helpful to explore several educational MOOs and choose one whose purpose and theme fit your needs. You will need to find out whether the MOO allows classes online and whether you need to request this in advance. In most MOOs you can simply send a MOOmail to the wizards with questions and the time for scheduled sessions you want (type '@wizards' or 'help wizard_list' to see names of administrators). It is also important to find out where in the MOO the administrators prefer classes to work. So be sure to check on this as well. Once you have been given permission and informed about the space you have been assigned, it is important to go there in advance to see how the room(s) work.

Many educational MOOs have specially programmed rooms for classes. In most cases the virtual classrooms duplicate a traditional RL classroom in many ways. You may find a clock above the door, a virtual blackboard, a teacher's desk, a work table, and various options for grouping students in small discussion groups. You may find more information about the room commands by typing 'help here.' The important thing is to understand how the room works before you bring students there. You may also be allowed to create your own classroom and/or virtual office in which to meet your students. That depends on the MOO building policy, but this is also easy to find out when you question the administrators about character policy, teaching policies, and scheduling.

As a final step in preparing yourself, it is always helpful if you have participated in some online discussions in the MOO you select. One of the most important aspects of using MOO in teaching is that you understand, and value, the nature of synchronous interaction in MOO. There is no better way to learn this than to participate. Most MOOs, especially educational MOOs, are communities of generous and friendly people. If you do not find answers to your questions from the help texts themselves, you can always ask someone online when you are there. You may page them privately with your questions or ask if you may join them to talk. In this way you will experience the MOO as your students will, and you will have understood the etiquette protocols that are so important to a productive and interesting MOO class. We will discuss MOO manners shortly, but there are still some preparation tips to discuss.

It will be important to know what class of player you and your students will be when they log on. As explained in the previous section of this book, some MOOs allow all players to have their own character, in which case you will need to request your own character in order to complete the preparation mentioned above, and you will need to instruct your students about how to request their own characters prior to logging on as a class. Some MOOs allow the teacher to have a permanent character, while the students log on each time as guests. You should understand the character policy of the MOO you select and request that your students read it as well. For example, some MOOs do not screen character requests and allow all players to be anonymous. That is, players are not identifiable other than by their chosen MOO name. This policy has its disadvantages. It is a problem if, for instance, one of your students is harassed by another player and cannot learn who the offending player is. Other MOOs require all players to be identifiable, allowing them to choose their own player name, but not to hide their email address if other players choose to use the 'finger' command to identify them. If students know from the outset that their identity is always available upon command, the offending behavior may be prevented. In any event, be sure to learn the policy of the MOO you select and instruct your students about the character policy and implications for future problems.

Finally, as a crucial aspect of preparing your students for their first MOO session, it is helpful to hold a general discussion of what they should expect when they log on, why you are using this space to conduct class, and what possibilities for enhancing their learning the MOO holds. You may want to open a discussion with questions from the students prior to logging on, allowing them a chance to openly articulate any concerns they may have. Of course, much of what you tell them can only be truly understood as they experience it online. It has always helped, in my experience, to contextualize this activity for the students in the ways I have described. If you want, it is also interesting to give a brief history of educational MOOs, explaining how they originated and how they are proving beneficial in educational environments (see the introduction and appendix).

Preparing Your Classroom or Lab

Let us assume you have permission from your department but do not know whether your classroom (if it is a computer-networked classroom) or lab has the necessary software to use MOOs. If you are using a lab, make sure the computing services director knows you are taking your classes to a

MOO and ask whether the computers have Internet access and Telnet capability. If you are using a computer-networked classroom, you can also request this information from the person who supervises the classroom. Either way, Telnet capability is a must for those rooms with Internet access. (If you are working with your own MOO, at the very least you must have locally networked machines. See chapter 5 for information about setting up your own MOO.) If you intend for your students to have permanent player status at the MOO you will be using, they will need to have email addresses in order to request characters at the MOO. So it is important to plan ahead and make sure all your students have email. (In some cases, you may be able to negotiate with the MOO to use one central email address for all your students, though this must be arranged in advance.)

Further, one of the most important components to a successful first MOO session (not to mention subsequent sessions) is for the classroom computers or lab to have a MOO client program.[2] A MOO client acts as an interface between raw Telnet and the MOO program itself. (Moo clients are described in chapter 3.) MOO clients for Macintoshes and PCs differ, and there are several varieties for each. As you have probably realized from your own participation in synchronous environments like MOOs, the stream of text entered by players is not controlled by the telnet program that gets you there. Without a MOO client, a participant typing something to say or emote will be interrupted by the text of another participant in the MOO. This can be very disconcerting for unexperienced players. Some MOOs offer ways to suspend the output from the MOO while you are typing, but a MOO client gives the best results. Many of these are freeware, and you may download them from any number of sites on the WWW.[3] If, for example, you are using TinyFugue, a program that works from UNIX, you may be connected to more than one MOO at a time, in addition to the cabability of shelling back to your UNIX account to open your email program, or to open other Telnet connections. If you use MUDDweller, a Macintosh client, you have a separate window for your output so that input from other players does not interfere with your text.

The important thing to remember is that clients make your classes go much more smoothly, and they will insure a good first experience for the new MOOer. Once you have the MOO client loaded on each computer, you are ready to test the program and instruct the students about its use. Most of these programs also allow you to set up a permanent file for the MOO you have chosen, so all that is necessary is to select that file and the students will go directly to the login screen of the MOO. Some teachers provide disks with the program on it and the file to that MOO, so students may access the MOO from home or other computers the same way they are used to doing in your class or lab. You may want to write up a brief set of instructions for using this program prior to your first MOO session as a class.

The First MOO Session

You may think all of this preparation will prevent chaos during your first class MOO session. Indeed, it will to some degree. The fact of the matter is that your students will still react in widely different ways to the MOO interactions and the text scrolling rapidly on their screens. Some will find it fun, some will be confused, some will be indifferent, and some will find it difficult, if not pointless. That is to be expected in a new and virtual environment. In my experience it is more productive in the long run to let the students explore and experiment, spoof and practice, provided it does not turn into anything inappropriate.

The best way to avoid that is to ask the students to read the 'help manners' text for the MOO (a brief text on netiquette called by different names depending on the MOO). A quick glance at the 'help' topics will give them the proper command. Once they have read the manners information, it is wise to stop the class and discuss the protocols for communication and movement, like knocking on doors before entering private offices or rooms, and refraining from inappropriate emoting or obscene language (just to name a few taboos). If there are still problems, you have capabilities of discipline at your disposal; we cover those later in the chapter.

Following a discussion of manners at the MOO, you are ready for your first activity. There are disputes among those who use MOOs about the wisdom of planning activities for first-time class MOO sessions. Some teachers prefer to let students explore at random, practice communication and movement, and so forth, before moving on to something focused. Others feel that the best avenue is always to have planned activities or assignments to keep students focused and to prevent play and idle class time. Both strategies are valuable; the best choice depends on the teacher, the type of class, the dynamics of the student interactions offline, and the goals of the MOO session itself. If, for example, the goal is to allow students to become familiar with the space before tackling an assignment or group discussion, then letting them explore and experiment is best. If the aim is some specific task, like peer editing of drafts of papers, then it is best to move them quickly into a virtual classroom on the MOO and teach them how to work with each other's drafts, or to discuss the drafts in small groups. The main thing to remember is that there is no one way to proceed; different practices work for different occasions and pedagogical styles.

If you prefer to begin with a focused activity, here are a few things to try. Prior to logging on, brief the students on the task of finding an object you have placed somewhere in the MOO, a kind of hide-and-seek. You could opt to place your own MOO character somewhere in the MOO and ask the students to find you. Tell them only enough to get them logged on, and let them know how to find the help texts, then set them loose. This works particularly well for traditional-aged undergraduates no matter what kind of course you are teaching. Most will enjoy the activity, though some may find it silly and boring. Instead of worrying about those students, use that to your advantage. Identify these students as fast as possible (you will know them by their comments on- and offline), then put them in charge of organizing (or rounding up) the rest of the class into some common space on the MOO. Then set up a recorder and record a discussion of the reactions of the students to the MOOspace. A good way to end that kind of first session is to give them an assignment for their next MOO activity. You could impress them by showing a MOO slide, or by writing on the virtual blackboard and ask them to 'look blackboard,' or prompt a recitable note with your assignment on it.

Another activity to try in your first class session is to ask students to describe their MOO character, set their gender, and give them an object to create as an assignment. This obviously works only if the MOO you are using allows students to have characters of their own. Even if it does not, most MOOs allow guest characters to describe themselves and set their gender. We will discuss these activities in the next section of the chapter.

Finally, it is important to allow students to discuss their reactions to the MOO experience. You will have to facilitate, of course, any discussion that tends to downplay the usefulness of the MOO in class, especially if those students who found it unappealing begin to dominate the conversation. See if you can solicit other comments from students who were obviously having a good experience, who caught on quickly, and who look forward to the next MOO session, or asked how to build their own space, and whether they may meet other students outside of class at the MOO. These are all indicators that the students found the experience helpful, fun, and educational. It is certainly important to allow students to voice their negative reactions, but you can use that to your advantage and ask them to log on to the MOO and set up a newsgroup or notice board where class reactions may be posted. By placing them in the position of making their reactions public and recorded, they will often change their opinions after having to learn for themselves how to create an object, post something, and then explain to the class how they achieved the assignment. Such an exercise gives them a sense of how beneficial the space can be, especially if their first experience in the MOO did not seem to lead to anything useful. The idea of play is an ironic notion that can be used to your advantage in such situations, but it must be finessed carefully.

Building and Describing

Most of the protocols and activities discussed so far have assumed that your classes have been held in MOOs administered outside your own institution, and that you have been limited somewhat by the policies of that particular MOO. If, however, you are taking your students to a MOO that allows students to have permanent player status (or has a temporary student player class), or better still, if you are using your own MOO (or one administered by your institution), then there are more activities available to you and your students. Specifically, I am speaking of your students contributing to the virtual MOO community by building and describing their own spaces and by creating objects that other players may see or use online. The value of allowing students to create their own rooms or offices depends on how invested you are in the MOO, that is, in how often you plan to hold class online, assign collaboration online, or hold office hours online.

Before I explain some of the creative ideas for student building, let's assume you are using a MOO that allows students to be guest players only. When you have prepared the class as outlined above, ask your students (after they read 'help manners') to set their gender and describe their character. Have the students type '@gender'; this is what they will see:

```
Your gender is currently neuter.
Your pronouns:
  it,it,its,its,itself,It,It,Its,Its,Itself
Available genders: neuter, male, female, either,
  Spivak, splat, plural, egotistical, royal, or 2nd
```

Students should then type, for example, '@gender female', or whatever they have chosen. Next, students should describe their character using the @describe command. If they first type 'help @describe', this is what they see:

```
Syntax: @describe <object> as <description>

Sets the description string of <object> to
  <description>. This is the string that is printed
```

```
    out whenever someone uses the 'look' command on
    <object>. To describe yourself, use 'me' as the
    <object>.

Example:
Cynthia types this:
  @describe me as "An interesting woman with a big smile
    on her face."
People who type 'look Cynthia' now see this:
  An interesting woman with a big smile on her face.
```

Many educational MOOs prefer students to use their real names, but descriptions may be creative, which gives the students some flexibility for exploring identity. You will want to make sure students have not been *too* creative, however, since their descriptions may be viewed at any time by other class members as well as other players not from your class or university who happen to be online at the same time.

Now let's assume your students are using a MOO that allows them to have "builder" status. Some MOOs require that players build offices and rooms that are connected to the MOO's public space; others do not. Thus, you should check the building policy at the MOO you use before you assign your students building activities. They may type 'help building' for instructions. An example might look like this:

```
help building
There are a number of commands available to players for
  building new parts of the MOO. Help on them is
  available under the following topics:

creation -- making, unmaking, and listing your rooms,
  exits, and other objects
topology -- making and listing the connections between
  rooms and exits
descriptions -- setting the names and descriptive
  texts for new objects
locking -- controlling use of and access to your
  objects
```

Other building information may also be located in help texts online, like the following note located in the Help Kiosk at Lingua MOO:

Housing Policy and Building Help
New players who have read the tutorials on creating rooms and objects may now go to the CoMOOnity and read the Note about creating their room. All registered players at Lingua can get a room in the CoMOOnity. Once inside the CoMOOnity, just type 'Newroom' and one will be created for you. Only one room per person can be connected to the coMOOnity. Your name will automatically be added to the Directory of Residents and visitors will be able to find you. Once your room is created, you have 3 options for the type of room you utilize. Follow the instructions of your generic room description to choose which parent room you want, then @chparent it as instructed. Then you may follow the instructions to describe your room and set seating details and other details of your choosing.

Still another helpful way to add descriptive information about your player may be found on certain MOOs that allow and encourage players to set their @interests. In Lingua MOO, for example, we place this information also in the Help Kiosk:

How to Set Your Interests
Each resident of Lingua should set your gender, describe yourself, and set your interests. The instructions for setting gender and describing yourself are in the help tutorials. We have added a feature that highlights one of Lingua's key purposes: a research interests feature. To set your interests, which is much the same as a simple description, type: @interests me is etc etc etc. This command enables players to do more than just look at another player, although the look command gives the player's description. The finger command will show the research interests information and is entered by typing 'finger player'. For example, if you want to see the broader set of information about Knekkebjoern, you would type 'finger knek'.

Providing an @interests description also informs other teachers and researchers who use the MOO about you and your specific interests, so don't forget to add your own and model the process for your students.

Recording Class Discussions, Archiving, MOOmail, and Newsgroups

There may be occasions when you wish to record and archive class discussions and other MOO conversations. You will need to check with the MOO administrators, but most educational MOOs offer several forms of recording devices. Instructions for creating these kinds of MOO objects vary from MOO to MOO since these are generic programmable objects. (You will find more information in chapter 4 about the use of educational recording objects.) If you choose to record your students' discussion groups, it is important that you make them aware that they are being recorded and give reasons. For example, at Lingua MOO we encourage teachers to use transcripts of group discussions as bases for analysis and further discussion. Each transcript is a mailable $note and can be emailed by the students to their email accounts. Not only do they preserve their discussions, the tapes may contain useful material for research in their writing assignments. This applies to all kinds of classes regardless of the topic of the course.

You may also want to create a $container for archiving the discussion transcripts so that students who do not have email accounts have access to them. Keep in mind, however, that transcripts take up considerable space, so once you no longer need them, please @recycle them. Another method of archiving discussions would be to put them on a web page and have your students access them that way, thereby freeing up the MOO space sooner.

It is important to consider the many ways that MOOs enhance RL classroom learning activities by adding (recordable) synchronous discussions, but don't forget that another aspect of MOO class activity is asynchronous. The MOO has its own internal MOOmail system that allows you to communicate with your students and them to communicate with each other. At some educational MOOs teachers may request special newsgroup lists for their classes, giving the students the ability to post messages to the whole class, or to their group, during the week between classes. This extends the notion of class beyond the "space" of the RL walls and the limitation of the RL class time. In addition, these newsgroups are another form of archiving discussions and student contributions to the class without having to set up a listserv through your academic-computing department.

Shared Projects, Tutoring, and Collaboration

Teachers often find that after they and their students are comfortable with basic MOO commands and activities, they want to explore more ways to use the MOO for productive classwork. Some MOOs offer various ways for students to share ownership of objects in order to collaborate more easily. For example, at Lingua MOO students may apply to the administrators to set up a multiple-owner project in which they identify several players who share ownership of a $note on which they are composing a shared document, or of another kind of object that requires textual work, like a "bot" (robot) programmed with keywords and actions.

Other kinds of projects that may result from more advanced building activity may involve a series of rooms, or matrix of rooms, that take players on a journey and that involve rich textual description and narrative, interesting ambient details, and special objects that players may activate or look at while exploring the rooms. Often students will become quite involved in their projects, even to the point of holding online conferences with each other about their work. In some instances, you may want to assign your best students to act as tutors for lower-division students. In the next chapter, Eric Crump describes what some online writing labs (OWLS) are doing with tutoring.

The MOO offers a rich virtual (and textual) environment for collaboration, and it does not require setting up projects for a group of players. One quick and easy way to work together is to record a discussion and then use that $note as a basis for a paper or presentation, editing it where necessary to eliminate the conversational nature of the narrative. Another way to collaborate is to @paste text from a word-processing file in another window right into the MOO, enabling the players in that room to see what you want to discuss, and to edit the text through subsequent pasting. Still another way would be to create MOO slides, or chunks of text, that you show to the players in your room or office, and they serve as a basis for discussion. Of course, if you are recording everything, then the transcript will reflect the pasted text and the slides as well, creating a rich collage of dialogue, straight text, and other snippets of text shown by MOO slide projectors. The possibilities are so varied and extensive that once you and your students learn how to achieve such collaborative building and creation, you will want to step up to more advanced skills like MOO programming, holding online events, and symposia.

Events, Advanced Projects, Teleseminars

There are unlimited possibilities for holding online events at educational MOOs (though you want to know the MOO's policy on such planned activities). Usually they are happy to help promote and organize events and help you select the most appropriate space in the MOO for holding the event. (In chapter 9 you will learn about the first-ever online dissertation defense held at Lingua MOO in July 1995.) Other possible events run the gamut from panels of speakers who debate on selected topics to open-mike poetry readings to staged performance events (see chap. 10).

If you decide to plan an event yourself, or allow your students to do so, get your plan ready and present it to the MOO administrators. Be sure to leave plenty of time for preparation, coordination, and promotion of the event (if you are inviting others to attend). Make certain you understand how the room works in which your event will take place. Rooms that are moderated are usually available, though you want to discuss the options with MOO administrators to understand the advantages and disadvantages of semimoderated and fully moderated rooms. Finally, stage a rehearsal with the invited speakers and/or students who organize the event. This is especially important if you have guest speakers who are unfamiliar with MOO. Send them beginner's guides to MOOing in advance and assure them you will help them throughout the process. In some instances when formal events are held, you might want to create objects such as plaques and programs for the full effect of RL events.

If your students have applied for an advance project that requires programming or extensive building, be sure you have permission from the MOO administrators. You may also want to check on the backup system of the MOO to insure that students do not lose valuable code or objects in a crash. If they will be building, make sure they have sufficient quota; if they do not, request more from the adminstrators. If they require shared ownership of objects, this also needs to be approved in advance. Much time will be lost if you or your students must wait on approvals at crucial stages in the project's development.

Another more advanced use of educational MOOs for teaching concerns the collaboration of more than one class in a teleseminar. Some educational MOOs have notice boards available for posting calls for participation to locate other teachers interested in planning a teleseminar. You may also want to post calls for participation to various listserv discussion lists that you know concern your teaching interest. Once you find another class and you have received permission from the MOO you want to use, you and the other teacher(s) should meet online to discuss the specifics of MOO collaboration among your students, and the possible overlap in theme, readings, and projects. You may want to set up a newsgroup at the MOO for the teleseminar and email instructions for subscribing to it to your students at the beginning of class. In addition, make sure your students have examined the basic MOO commands and 'manners' help texts. A valuable source of research in collaborative learning and online communities, a teleseminar web site and/or archive of class logs and syllabi will serve to preserve the experiences of your students in an invaluable mode of education.[4]

Conferencing, Research, and Evaluation of MOO Activity

In addition to the many ways your students may benefit from the MOO activities you assign them, it is important to remember that in most cases the MOO is not a substitute for the physical classroom space and the real/actual activities that take place during class time. Many teachers are experimenting with varying levels of involvement with educational MOOs, and most are finding that not only does it stimulate their students' learning, it enhances their own teaching and research.

Online conferencing is often more convenient for both students and teachers. Some teachers like to hold office hours in both the RL office and online in their MOO office. They may also request that their students conference with each other online, as in team conferencing, peer editing, and other collaborative activities mentioned earlier.

Teachers are also finding that the MOO is not just a place to hold classes, or meet individually with their students—it is a place to meet colleagues, to meet new people with shared research interests, and to attend meetings of other teachers using educational MOOs.[5] There are weekly meetings at Connections MOO (called Tuesday Cafe) on topics primarily related to teaching writing in computer-networked environments, and meetings at Lingua MOO (called C-FEST) on topics related to delivery and professional issues. Most educational MOOs are now also setting up their own WWW interface so that teachers who use MOOs may benefit from the resources available at any number of MOOs and/or their WWW sites (see the appendix for those MOOs with WWW addresses). In addition to valuable practical and technical information about teaching with MOOs, some of these sites also contain links to published research about teaching in virtual learning environments. The topics range from identity politics to intellectual-property issues to cyber frontier law and order.[6]

One of the most interesting issues, in my view, about using educational MOOs in teaching is how to evaluate MOO participation and textual products. One researcher of virtual learning environments, Peg Syverson of University of Texas at Austin, has formulated a model for evaluation of "activity" based on how some grade schools evaluate the activities of children. According to Syverson (who teaches graduate- and undergraduate-level writing and rhetoric), higher-education teachers need to understand how to shift from product-oriented to activity-oriented evaluation, or at least to some combination of the two. You may check out her course description, which outlines how her students can expect to be evaluated, at http://www.cwrl.utexas.edu/~syverson/basicinfo/evaluation.html.

One thing is certain, the use of educational MOOs opens up a wealth of new learning opportunities and new ways of relating to our colleagues and to our students. The issue of assessment has often been raised as one of the two best reasons for *not* using MOOs. The other one relates to misbehavior and harassment. The MOO is a community as real as any actual community, and social interactions in real communities carry social consequences. Thus, most educational and professional MOOs have strict guidelines about inappropriate behavior. In the next section, I address misconduct on MOOs. Make no mistake, however: neither of these reasons create insurmountable problems. In my experience, those who adminster MOOs and who teach in them are some of the most creative and generous people in academia. Together we have sought to right wrongs, to write rules, and to push the envelope of this cutting-edge learning environment.

How to Handle Discipline Problems

Many inaccuracies about the difficulties of student conduct problems on MOOs circulate in the wake of the much-hyped cyberrape that occurred on LambdaMOO (see chapter 2, about LambdaMOO, and chapters 12 and 13, which deal in part with the issue). Poor conduct is nothing to take lightly in any classroom or other space, whether actual or virtual, and must be overseen with care, especially if you take your students to a large, open educational MOO, where they are likely to encounter players from all over the world. As I mentioned in the preparation section of this chapter, you cannot prepare your students *too much* on this issue. A time may arise in which you or your students must take action against harassment—or worse.

Each MOO has its own set of procedures of discipline, but most have information that you and your students may want to discuss to prevent conduct that would warrant disciplinary action (see chaps. 3 and 5). Let me stress that each incident requires a different response, just as in 'real' life. As long as you have your eyes open and your students are prepared, the chance of encountering inappropriate behavior is minimal, and the knowledge about what to do if it happens is readily available.

I feel strongly that the benefits of using educational MOOs in teaching heavily outweigh the occasional problems. I hope that this chapter (along with the others in this book) helps you to plan properly and sufficiently and to understand and anticipate some of the problems you may encounter. Whether you decide to create your own MOO or use an established educational MOO, you should have unlimited resources at your fingertips in this exciting venture. Your students may perhaps never see all the planning and behind-the-scenes work you do to make this experience beneficial, but that applies to *all* teaching, yes?[7]

Notes

1. For a brief and helpful handout on preparation, see Tracy Gardner's tipsheet at http://www.daedalus.com/4cmoo.html.
2. Nick Carbone has written a useful handout on MOO clients that may be accessed at http://www.daedalus.com/net/telnet.html.
3. See the TECFA MOO educational technology pages, especially the section on MUD and MOO clients (http://tecfa.unige.ch/edu-comp/WWW-VL/eduVR-page.html#Clients) for links to various FTP sites where clients are available for downloading (both Mac and PC versions). See also this comprehensive web site for information about all kinds of MUD and MOO clients (http://www.math.okstate.edu/~jds/mudfaq-p2.html).
4. For an example of this kind of teleseminar, see the link to the COLLAB teleseminar (http://wwwpub.utdallas.edu/~atrue/COLLAB/) held in the fall of 1995 off the Lingua MOO Archive and Resource Page (http://lingua.utdallas.edu/archive.html).
5. One educational MOO group has organized itself through the GNA (Globewide Network Academy) network of MOOs, and they sponsor occasional meetings announced on the "moo-ed" listserv, rotating their meetings in different time zones and at different educational MOOs for maximum exposure to other MOOs and other researchers in this rapidly growing field. For complete information about how to join this group, see their web site, maintained by Daniel Schneider of TECFA MOO (http://tecfa.unige.ch/moo/edmoo/).
6. For an excellent example of research on discipline issues on MOOs, see Charles Stivale's 1995 MLA presentation, "Help Manners: Frontier Tales of Two MOOs," archived on Lingua MOO at http://wwwpub.utdallas.edu/~cynthiah/lingua_archive/help_manners.html.
7. For other resources for teaching with MOO (online articles, syllabi, assignments, etc) see our Lingua MOO Archive and Resource page, accessible from our home page at http://lingua.utdallas.edu; the TECFA MOO Educational Technology web site at http://tecfa.unige.ch/edu-comp/WWW-VL/eduVR-

page.html#Contents (see Teaching IN and ABOUT Cyberspace links); and our CoverWeb article on teaching and research at Lingua MOO in the electronic journal *Kairos* 1, no. 2 (summer 1996) at http://english.ttu.edu/kairos/1.2/index.html. Look for the CoverWeb link, then click on the first article, "Lingua Unlimited" (http://wwwpub.utdallas.edu/~cynthiah/start.html).

8

At Home in the MUD

Writing Centers Learn to Wallow

Eric Crump

Writing centers often build their sense of identity and worth based both on what distinguishes them from writing classrooms and on how they complement the classroom. Network technology, particularly a tool like MUD, profoundly changes the landscape of education and how we write our way around the new terrain. With these changes in writing, the identity terrain for writing centers changes as well, and the time may come—soon—when writing centers will want to escape the orbit of the classroom and build their sense of identity and worth based more on what distinguishes them from their own current form. They may, in other words, want to make something of an evolutionary leap. In their current form, writing centers are often primarily sources of aid for student writers struggling to produce texts in academic discourse specifically to meet academic requirements. They may begin serving as sites for virtual communities of writers, as places where a buzzing social life *constituted* in writing will thrive. And if they do, the tools at hand that are best suited for evolutionary launchings—in the immediate future anyway—are MUDs.

This chapter portrays some early forays of writing-center work into MUD environs. It also includes consideration of the implications that this terrain, recently discovered by writing centers, may have on how writing centers work. It is important to move quickly past the almost-inevitable objections to most enthusiastic portrayals of network environments, namely, that face-to-face interaction cannot be replaced. *Replaced* is not likely (in the short term, anyway) or even desirable, and fretting about it diverts our attention from the more interesting and more consequential questions of just what the relationship might be between traditional writing centers and new online writing centers, between writing centers and writing classes, between asynchronous print and electronic communication and synchronous conversation in writing.

Here is a progression of how writing-center specialists may apprehend MUDs:

At first glance: An interesting tool with vague possibilities, something a bit futuristic and geeky and odd. Not our bag. Come back next year.

At second glance: An interesting tool that can support something like traditional, one-to-one tutoring, but over distance! Might have some applicability, after all. How do we get one?
At third glance: A new dimension. A world to inhabit. A new world written into existence using familiar words. MOOing is *living* with writers, living in writing, a whole new kind of thing. Writing centers become residual artifacts of print culture and give way to virtual writing environments, the space for living, conversing writers.

There are a number of reasons the match between writing centers and MUDs is a good one (though ultimately disturbingly and disruptively radical), but at the heart of the matter is the "oralish" nature of synchronous written communication. The main strength of writing centers may be in a common structure that typically puts writers and tutors in dialogue with each other. Experienced tutors often attempt to exploit students' natural oral fluency in order to help them better express themselves in writing. The main problem with writing centers is this same tendency toward oral dialogue. Yes, the strength and the weakness are, as often happens, the same. Writers and tutors talk and talk, often helping students make wonderful progress toward understanding the subject at hand, toward developing the ideas and arguments brought to bear upon that subject, but if no ink or electrons are applied immediately to the process, a problematic gap is opened in the project, a gap between engagement with ideas and recording ideas. What is often lost, in the time between the conversation about and the inscription, is the rich immersion in detail and nuance, the *intellectual energy of the moment.* When that becomes a problem, it is time to look for a nearby MUD to jump into.

That leap is the "second glance" and looks, at first, like a sort of bridge from current practice to current practice online. It is actually a leap into a new dimension, but we will get to that later. What MUDs provide writers and tutors that is immediately appealing and connected to current practice is a chance to be immersed in conversation, to generate energy, *and* to capture the moment. That is a powerful combination that cannot be constructed in most face-to-face (f2f) writing-center situations. In moments when I temporarily forget the powerful preferences many people have for f2f interaction—in spite of its inadequacies as a situation for writers and writing—I am amazed that writing-center specialists aren't, en masse, clamoring for MUDs, demanding MUDs, migrating MUDward. Of course, f2f can be just the thing in some situations, but there are ancient biases at work here that only subsequent generations, socialized to see virtual environments as routine and normal, will shed. Otherwise, MUDs should seem almost too good to be true for writing teachers and tutors.

Second Glance in Action

To illustrate, I would like to share some excerpts from logs of MUD sessions I had with a composition student during the late fall 1995 semester. "Tomas" had been referred by his teacher, who was as frustrated as Tomas was by his inability to write and revise up to expectations, in spite of extraordinary effort on his part. Tomas and I never met f2f, but he sent me drafts of his papers via email, and we met five times on a local educational MOO, ZooMOO. I would like to influence your reading of the exchanges between Tomas and me by noting that when colleagues ask me how writing teachers or tutors might use MUDs, I often hold this case up as an example of the potential such environments have for helping students with their writing. It seems like such shining example. Tomas seemed to make good progress. On an evaluation form for the Online Writery (an environment for writers that ZooMOO is part of), Tomas said, "[I] completely owe the saving of my grade in English 20 on computers to the enormous amount of help and aid from eric crump." And his teacher, in a private email note to me, said, "His essay was 100% better than anything else he had written." Wonderful experience? Yes. The ultimate example of how MUDs and writing centers converge? Not a chance. Lest you think I am bragging here, all is not as rosy as it seems, and I will discuss later why I think the dialogues to follow represent only a modest advance in the way of real writing-center evolution. MUDs are going to affect us more decisively than this. (Please excuse typographic errors in the following exchanges. Real-time written conversations are rarely error free, even among accomplished writers. Rather than edit the logs to fit print conventions of correctness, I chose to leave the errors that did not seem to unduly muddy the meaning so as to preserve the flavor of the discussion.)

> [November 7]
> Tomas says, "i'm in really bad shape in [this] class, so [my teacher] is giving me another chance. she just thinks i didn't revise it good enough. i honestly thought i revised the WHOLE paper, except for some quotes" (Tomas)[1]

Tomas comes to me for help out of desperation. His frustration emerges with some regularity throughout our discussions. He works *hard* and even tries to pay attention to the clues, spoken and unspoken, but cannot seem to decode his teacher's expectations. Desperation is not the best state in which to write well for most people. I get the impression that Tomas is nearly paralyzed by it. He does not know how to proceed, even at a very practical level. He does not know what words will work, what magic rearrangement will satisfy his teacher's apparently obsessive and inscrutable demands for revision. His situation is not uncommon in writing centers. The desperate make up a significant minority of the students who visit mine. Our job, in the f2f writing center, is to help writers discover better revision strategies. We coax new versions. We ask questions that we hope will turn on lightbulbs. We try to provide an additional rationale for the teacher's inexplicable obsessions. We talk and point to the paper and make notes, and as students head down the hall after each session, we hope they will manage to churn out something good, something they can be proud of, something their teachers will admire and reward with a nice grade.

That's what I did.

```
[November 10]
Tomas says, "i think [my teacher's] big problem with
  me, is that she doesn't think i revise my papers good
  enough, or to her standards or wishes"
Tomas says, "even though, as you could see, i
  completely revised it"
Eric says, "hmmm. well, it could be that what she means
  by revision and what you mean are different. could be
  a misunderstanding of sorts"
Tomas says, "it was a misunderstanding the first time,
  but the past two, i really tried and thought i
  completely revised them"
Eric says, "you moved things around and changed things,
  but the third draft is still essentially the same
  paper. many of the passages from the first draft still
  exist, whole, in the third"
Eric says, "now, that's a perfectly acceptable way to
  revise for many teachers."
Tomas asks, "no, isn't it noticeable, that i changed my
  whole thesis, my way of LOOKING at the issue?"
Eric says, "well, I see the shift, yes."
Eric says, "we may be talking a matter of degree here.
  you made a shift and she wanted an earthquake"
```

Eric says, "here's something I've asked students to do when they come to the writing lab and are kinda stuck on some part of a paper . . ."
Eric says, "I turn over the draft, get out some scratch paper, and tell them to write a new version of the same idea."
Eric says, "I think you *did* make a good effort at real revision. I'm just thinking maybe [your teacher] was looking for something *even more*"
Eric says, "by the way, the students who did the start-from-scratch thing in the lab often admitted that it helped, but they also groaned when I suggested it"
Eric says, "it's hard to give up those words we've worked so hard to create"
Eric exclaims, "I know. I'd hate to do that, too!"
Tomas says, "so i have to start from scratch"
Eric says, "it's a possible strategy. you don't *have* to."
Eric says, "I'm not your boss, ya know! just trying to think of ways to tackle this problem"
Eric says, "maybe we can think of another approach, too. nice to have options, I always figure"
Tomas says, "i'll do it, but i have a big paper to write in another class, and an even bigger one in english, i'm wondering when i'll be able to do it. i think it's seems obvious that the only real way of doing this is to rewrite it"
Eric says, "what I was thinking [a] new intro might be is some explanation about *your* answer to the questions. maybe 100-150 words about how you make decisions about what's good, bad, or true"
Eric says, "then get into your sources and how they support or depart from your stance"
Tomas says, "well do you think she likes me talking about myself"
Eric says, "well, the way I read the assignment, you have to describe your own method of answer the questions. you may not really be focusing on yourself so much as on your methods"
Eric says, "give it a go, anyway. feel free to send me

what you come up with, even if it's rough. in fact, rough is OK. I'd worry about crafting it later"

This exchange looks quite similar to what might have transpired if Tomas and I had been in the writing center, his marked-up draft sitting on the table between us, his flimsy, scarred albatross. I try to be pals with him. I talk too much. He's so intimidated by his teacher that he can't think straight, and I hope that if he thinks of me as a different sort of creature altogether, a friend rather than an implacable judge, he will relax and let his brain and hand begin to work at the revision challenge. After this exchange, I am cautiously hopeful. He seems willing to take my suggestions.

The main advantage that holding this session on the MUD offers, a small but possibly significant thing, is capturability. Tomas is new to MUDding, but I keep logs as a matter of routine, so I email him a copy. He now has a record of our discussion, of the suggestions I made, of his own answers to my questions. It is a previously unavailable resource. How he will use it is another question. But it's good that he has it, I think.

[November 28]

Eric says, "we might want to look at a few sentence-level things. when you get your ideas down it helps to get them expressed as well and clearly as you can. here's one [sentence from the draft].

Personally, as is for many people, this answer, yes, comes from personal convictions: my values installed from my family life, my religious upbringing, and my own personal views on whatever it is I am making a decision on.

Eric asks, "that's kind of a difficult sentence to wade through. I wonder if we could capture the same idea but phrase it differently?"

Eric asks, "I mean, if you were just to tell me now what you were trying to get across there, what would you say, just off the top of your head?"

Tomas says, "should i say 'the answer yes comes from personal convictions: values installed from family life, religious upbringing, and personal views on whatever it is one is making a decision on' but if i say that, it eliminates the personal factor"

Eric says, "I do like the way you phrased it better . . ."

Tomas says, "so i should say 'my' before the colon, and not in what's after the colon"

Eric says, "sure. that would work"

Here's an example that underscores the importance of that discussion log as a resource. Tomas's new version is more graceful, better than what's in the draft. But rather than having to re-create that better version, Tomas has it in hand. He can paste it right into his next draft if he chooses. This, it seems to me, is the great "second glance" advantage of MUDs for writing center work. The environment lets us quite easily do our bit as writers' allies. We can engage in the kind of informal and exploratory conversation that is our standard approach in the f2f writing center. And since we are just folks, just sitting around talking, students can often get past the intimidating influence of the page (I often think the $8\frac{1}{2} \times 11$ piece of paper appears to some students like some sort of white looking glass, and just on the other side is the Red Queen eternally reminding them: "Always speak the truth—think before you speak—and write it down afterwards") and just talk. We're "just talking" in writing, though. Natural oral fluency, usually ephemeral, is now capturable. If we use MUDs only as a venue for students who think they cannot write to prove themselves wrong, that alone might make them worth using.

I feel good about this session. It is not until I review the logs later that I notice an all-too-common pattern to the discussion. Tomas is more at home on the MUD than on the page, and he seems more comfortable with me than with his teacher, but he is still very dependent on my permission and approval before making a move with his text. He seems to see me as a kinder, gentler teacher—but an Authority, nevertheless, whom he must appease before his writing will be considered acceptable.

[December 5]

Tomas says, "i'm just really hurting for that extra whatever that will help me. she's also said that if the next version isn't to her likings, she'll immediately stop reading it and give a 0"

Eric says, "wow"

Tomas says, "her own words"

Tomas says, "actually 'if it isn't revised as she wants' "
Tomas says, "i'm scared"
Tomas says, "from my past experiences"

[December 6]
Tomas says, "what about the conclusion"
Tomas says, "not just a summary"
Eric says, "hmmm. could be a way to make this particular case fit in a larger context."
Tomas says, "sounds good, a good idea for the whole, or better of, but what"
Eric asks, "that is, skinheads are a case. what are they a case of? what's the bigger picture?"
Tomas asks, "public trust in media portrayals?"
Eric says, "there ya go. that might be the ticket."
Eric says, "we start with reference to the media portrayal, but that's sort of an undercurrent in the paper, which focuses on refuting a particular media image"
Tomas says, "let me type one up real quick, hold on"
Eric says, "the conclusion might talk about the larger issue of media images and how much we depend on them, how that affects our view of others, etc."
Eric says, "ok"
Tomas says, "Skinheads are just one instance, in where we, the public depend on media images for the way we view others. It can be seen that the media image of skinheads exists, but only for some individuals. Media images (advertising, and film in particular) have given us poor, distorted images of everyone, including women, homosexuals, and many others. We can see that the media is something to not rely on, for something that you should do with your own perceptions, feelings, and opinions, for any individual."
Eric exclaims, "yes!"
Eric says, "nice"
Eric says, "I think we're getting there, my friend"
Tomas asks, "not too repetitive?"
Eric says, "I don't think so."

This example represents progress, for this student, in revising a particular text and perhaps in writing skill and confidence. But pedagogically, it's a baby step, perhaps more lateral than forward. I'm tempted to cheer briefly and then say, "So what?" The patterns of interaction and the assumptions about authority that inform my exchanges with Tomas suggest that we've merely migrated from the traditional writing center to a new venue. That's an OK thing to do. I don't want to disparage the move. The benefits of being able to interact with writers at a distance means we have new opportunities to help them at their point of need, when they are neck-deep in the writing process, rather than at a deferred point, when particular challenges are less immediate, sometimes only partially re-created. And of course, the ability to capture oral eloquence in text is no small advantage. But I think we are only at the beginning, here, and the real role of MUDs, and similar tools, proceeds from here.

Third Glance: Writing Centers Become Writing Communities

Writing centers typically offer a wide range of services to writers, some of them reflected in the publicity fliers they produce: Come to us for help with organization! grammar! brainstorming! mechanics! citing sources! etc.! There remains a perception, justifiably so, that the "customers" of the writing center want textual triage, and that is what many writing centers feel compelled to advertise, however that might depart from the prevailing pedagogical stance (i.e., that tutoring writing is a process, not a quick fix). Other aspects of writing centers are reflected in the literature of the field: how writing centers can contribute to students' development as writers, to their written fluency, to their confidence and competence with academic discourse, to their relationship with various knowledge bases and intellectual communities. Even so, the actual day-to-day work in many writing centers is held in the gravitational field of texts—almost always monological texts—and, in spite of intentions to become something else, writing centers remain places for the treatment of texts rather than the congregations of writers.

What MUDs provide is an opportunity (not a magical solution) to complicate our relationship with texts, maybe even productively disrupt that relationship. MUDs are often described as "text-based virtual environments," and that is accurate, but in terms of their role for language and literacy education, MUDs might better be thought of as conversation-based virtual environments. Unlike texts we create for the page, almost all texts in a MUD are implicated in the immediate, social, local phenomenon commonly know as communities. This capability may serve as a bridge from the writing center (or at least its literature) to the virtual writing environments we create when we move into MUDland. Although there are some notable exceptions, most f2f writing centers are a place to talk *about* intellectual communities. They often serve as interpretation centers for students struggling to insinuate themselves into academic discourse communities. They rarely serve as places where communities actually form and flourish. Or if they do, as in those exceptional places where students actually *hang out* rather than only popping in periodically for help with writing problems, they do so on such a small scale as to be practically insignificant as an influence on the learning environment of their institutions or of education in general. I hope the migration of writing center work to MUDs will profoundly change that situation. As we make that move, we have to reconsider our traditional practices and attitudes, shifting from a print-based textual orientation that, current theorizing notwithstanding, still dominates our practice and our relationship with writers to a net-based orientation that foregrounds the chaotic conversations that serve as such fertile soil for community formation.

In other words, we are in a different business now, and MUDs are a more amenable environment than writing centers for doing this business. They hold incredible potential for hosting communities, and communities are better learning environments than classes or even current writing centers. Joseph Saling and Kelly Cook-McEachern make that point well in "Necessary Communities and Fluid Apprenticeships: The Potential for Writing Centers."

> It would be nice if writing centers were places where writers talk to writers. Unfortunately too often writing centers resemble mini-academies. The hierarchical structure of American education is reflected in the division of power among writing center directors, professional writing consultants, graduate assistants and peer tutors. At the bottom of the strata are students. This authoritarian structure prevents true dialogue between writers. The labels and their connotations are deeply rooted in people's minds and it is very difficult, no matter how democratic everyone tries to be, to dissolve these barriers for the establishment of writing and learning communities. (1)

Hierarchies may be inevitable (as some claim), but they need not dominate the educational landscape. Because they are control structures that constrain the distribution of power, authority, and responsibility vertically, the higher you go, the more highly concentrated is the power, authority, and responsibility. The lower you go, the more you are the object of those forces, being acted upon rather than acting, being required to do things rather than initiating action. That means that students, as the lowest strata in educational hierarchies, generally are required to acquire knowledge and skill. They are not allowed the freedom to explore their own interests, make their own mistakes, and learn from participation in a community of colleagues. As Saling and Cook-McEachern say:

> Our goal as educators in a democratic society should be to create the environment necessary to foster true democracy and to enable the students who work with us to participate in that democracy. We should be makers and enablers of democratic communities, and in fact, exist as an amalgam of such communities.
>
> These communities have long existed on our college campuses. They are called departments. They are called interdepartmental committees. They are teams of researchers and educators who look for new knowledge or for new ways to converse about knowledge. These communities consist of faculty, graduate students, and, in the best institutions, administrators. But few academic institutions in America have any more than a handful of undergraduate students who are active members of these communities. . . . The effect of isolating students by focusing attention on them as if they were the product of teaching is to systematically exclude them from these already existing communities. Unfortunately for students, it's within these communities where true learning takes place. (2)

Steve North and many other writing-center scholars have claimed that writing centers are, or should be, havens from or alternatives to the hierarchies of the classroom, places where the writer matters more than the writing. But to a great degree, writing centers do not escape those hierarchies but merely configure them differently. To a great degree, those of us who are exploring online environments for writers are surrounded and shaped by the hierarchical establishment. So if we simply state that we are about creating communities for writers, we are not ahead of our f2f writing-center cousins who assert the same thing. To really create an environment in which communities can flourish, we have to *act* differently, in ways that are sometimes uncomfortable and sometimes in opposition to the powerful status quos around us.

Community Is the Opposite of Bureaucracy

Community is guided by mutual need and reciprocity. Authority is distributed. Bureaucracies tend to use authority to impose control. In communities, consensus and dissensus coexist, mingle, oscillate. In bureaucracies, consensus is mandated, dissensus is related to an appeal process, funneled into back alleys where it cannot upset the apple cart. But it is not *necessary* in all cases to express criticism privately. That does not imply positive news to the exclusion of negative news or disagreement. It means delivering the news with respect and requisite courtesy. It is possible to be angry with someone and still interact with them respectfully and productively. That's part of collegiality. That doesn't often happen in hierarchic, bureaucratic settings.

Collegiality for all—that is a radical notion, even in writing centers. It means disrespecting certain boundaries and decorums that tend to separate us (tutors) from them (writers). I am hoping that we slowly blur the distinction between staff and client.

> When students become colleagues, they are able to join us in the real work of the writing center. We can ask them for their input and then implement it. As peers, they don't need any special certification to participate—their grades don't matter. It is also easier to be honest with colleagues. . . . we can trust that the reaction from a colleague will be less defensive and more active than the reaction from a student who needs not just our judgment but our approval as well. (Saling and Cook-McEachern 6)

When I think of Tomas, his anxiety about writing exacerbated by his frustration, resentment, and confusion about seemingly unattainable standards set up by a teacher he felt was not only sitting in judgment of him, but who he suspected was withholding the "answers" to this baffling writing puzzle, then I think of the kind of relationship I've seen develop between teachers and students on MUDs. Here, for instance, is an excerpt from a MUD session that took place early in the semester, not long after I introduced the class to the local MUD. We'd gotten into a discussion about grades, with a nice mix of views expressed, some agreeing with my anti-grade stance, some questioning it. Some just goofing around.

```
Zimbo says, "Jasmine is really hung up on the
  alphabetical iron maiden . . ."
Eric says, "like Zimbo notes, grades represent
  evaluation that follows a fairly narrow and rigid
  path: teacher—>students, where teacher sets all
  criteria, students perform, teacher judges
  performance"
Jasmine asks, "What is that supposed to mean, Zimbo?"
```

cornfed says, "grades act as a hurdle or barrier that we all must climb and deal with because you can't get into good schools and you can't get any free rides w/ out (except jocks)"
Eric says, "alphabetical iron maiden? hah! I love it"
Auggie asks, "Did anyone actually listen to Iron Maiden?"
Zimbo says, "It's pavlovian learning . . . I'm not too fond of it."
Auggie says, "Oh no, Zimbo is getting historical on us"
Eric used to listen to iron butterfly when he was a kid, but not iron maiden
Jasmine asks, "You all act like grades are the enemy - what about people who really work for them and deserve tham?"
Zimbo tries to imagine Eric jamming to Iron Butterfly . . .
Eric says, "I figure evaluation and assessment are important. everybody wants to know whether what they are doing is good, right? whether what they say and write is having any effect on others"
Zimbo puts the thought right out of his mind.
Eric exclaims, "ina god a da vida!"
Larry likes calling radio stations and asking them to play the long version of inigodadivida
Eric says, "(or however you spell that)"
Jasmine exclaims, "Vindi Vidi Vici!"
Zimbo asks, "I work my hump off for grades I detest, Jasmine? Does that mean I deserve to fail, because the system bites and I don't like being chewed?"
Auggie says, "Zimbo, they're all out to get you"
Larry says, "ERIC, is it true you used to be an olympic weightlifter until you were caught using steroids"
Eric tries to look tyrannical
Larry asks, "just a little?"
Eric growls a bit
Auggie says, "I quiver in fear of eric"
Jasmine says, "You're doing good, Eric"
Zimbo finds the violence inherent in the system.
Jasmine says, "ooo . . . violence"

> Larry asks, "eric were you a cheerleader for the LA LACERS back in the '80's?" (Composition Class)

In spite of efforts over the years to put students at ease in the classroom, I have never seen students drop their guard enough to actually risk making fun of me (to my face). Even though my face was right in the same room with these students, as soon as we entered the MUD, we were in a different world, and they knew the rules had changed. The easy informality that seems to come naturally in the MUD contributes, I think, to an environment nearly free of fear. Teachers are often satisfied with forcing students to perform, and fear is an effective tool if mere obedience is the goal, but student learning is never served by fear. As John Holt says in *How Children Fail,* "The scared fighter may be the best fighter, but the scared learner is always a poor learner" (49).

For online writing centers, hoping to discover how network computing tools might help them create the realms in which student writers can develop and grow, MUDs are rich with possibilities. As writing begins to assume the shape of new technologies, dialogic forms will begin to displace monologic forms that thrived in print. Conversation is the stuff from which communities grow, and as Saling and McEachern note, communities are the sites where learning thrives. What online writing centers might aspire to become are "third places," Ray Oldenburg's term for "places [where] conversation is the primary activity and the major vehicle for the display and appreciation of human personality and individuality" (42). The best illustration might come from Richard Goodwin:

> A thousand minds, a thousand arguments; a lively intermingling of questions, problems, news of the latest happening, jokes; an inexhaustible play of language and thought, a vibrant curiosity; the changeable temper of a thousand spirits by whom every object of discussion is broken into an infinity of sense and significations—all these spring into being, and then are spent. And this is the pleasure of the Florentine public. (Qtd in Oldenberg 27)

And perhaps it will be the pleasure of online environments for writers and writing teachers. And perhaps MUDs will be the tool that provides the opportunity to create such places. Certainly writing "instruction" will not look anything like what it does now, in boxy classrooms or their writing-center cousins, but the development of proficiency and fluency with language will happen, nevertheless, and it will happen *in writing.*

Note

1. This MUD-based conversation and all subsequent conversations with students are included here without their explicit permission. However, all names, including names that were already pseudonyms (except the author's) have been changed, and all participants were advised at the time that the session was being recorded.

Works Cited

Composition Class. MUD-based conversation between members of a first-year composition class and Eric Crump. ZooMOO: moo.missouri.edu 8888. September 22, 1995. http://www.missouri.edu/~wleric/classlog.html.

Holt, John. *How Children Fail.* New York: Pitman Publishing, 1964.

Oldenburg, Ray. *The Great Good Place: Cafes, Coffee Shops, Community Centers, Beauty Parlors, General Stores, Bars, Hangouts and How They Get You Through the Day.* New York: Paragon House, 1989.

Saling, Joseph, and Kelly Cook-McEachern. "Necessary Communities and Fluid Apprenticeships: The Potential for Writing Centers." Unpublished essay.

9

Defending Your Life in MOOspace

A Report from the Electronic Edge

Dene Grigar and John F. Barber

On July 25, 1995, Dene Grigar, then a doctoral candidate at the University of Texas at Dallas (UTD), conducted the public, oral portion of her dissertation defense in the auditorium of Lingua MOO, a text-based electronic environment located at UTD and designed for education and research.

As far as is known, this was the first documented online dissertation defense, and it encouraged outside scholars and other interested individuals, in addition to the traditional dissertation committee members, to question a doctoral candidate and participate in a discussion concerning the candidate's work.

Several people worked together organizing the first online dissertation defense and assisting during the proceedings. Many other colleagues contributed to the success of the event by commenting on Dene's work prior to or following it via email and MOO-based communications. The collaborative nature of this undertaking invites further collaboration in its evaluation; therefore, this report reflects the thoughts and ideas of many other scholars who participated in and contributed to the event.

Using metanarrative and ideographic perspective, the authors review the reasons for conducting this dissertation defense online, describe how the event was organized and presented, and discuss the implications that emerge, as well as the problems that arose. Finally, they make recommendations to others who may also wish to negotiate the electronic edge while defending their academic lives in MOOspace.

It should be noted that email messages and MOOlog excerpts are included in this chapter as they were originally written. The authors have not changed or corrected grammar, spelling, mechanics problems, or idiomatic usages of language—American or otherwise—except where inaccuracies hamper concise communication. This is not meant to embarrass the authors of these texts but rather to preserve the informal, text-based orality of their thoughts, as well as the archival nature of these texts.

The Event

Sean quietly enters.
Pat finds a seat in Auditorium.
JG enters the room.
Mindi waves at Dene.
Mindi senses Dene is looking for her in the Auditorium.
Dene pages, "Thanks. Thanks for coming."

It wasn't until I saw all of the people teleporting into LinguaMOO on the afternoon of July 25 that I realized for the first time the magnitude of what I was about to do. True, my collaborators and I had sent out invitations to all of the listservs that we belonged to—Megabyte University, Classics, Humanist, Alliance for Computers and Writing, and others. And true, there had been much talk generated on these listservs concerning this event. But I had been so busy with the presentation I was about to make—not to mention preparing for the questions I was going to be asked—that it had not yet occurred to me what I had been planning to do for the last few months would be one, risky, and two, difficult.

Date: Sun, 16 Jul 1995
From: Dene Grigar
To: Multiple recipients of list
<mbu-l@unicorn.acs.ttu.edu>
Subject: On-Line Defense

Invitation to an On-Line Dissertation Defense
"Penelopeia: The Making of Penelope in Homer's Story and Beyond"

On July 25th at 4pm (CST) I will be defending my dissertation on-line via Lingua MOO, a virtual . . . space located at the University of Texas at Dallas. My study, "Penelopeia: The Making of Penelope in Homer's Story and Beyond," traces the artistic response to Penelope in literature, the visual arts, and music, from the Middle Ages onward.

```
This will be the first defense held on-line, and my
collaborators, Cynthia Haynes (UTD) and Jan Rune Holmevik
(Institute for Studies in Research in Higher Education;
Oslo, (Norway), and I invite anyone who is interested to
attend. . . .

Participating in the panel discussion at my defense will
be, of course, my committee . . . as well as a few scholars
who are currently involved in research on Penelope. Thus
far, Patricia Moyer (UNC) and Katie Gilchrist (Oxford)
have both agreed to participate. Jeff Galin (U of
Pittsburgh), John Barber (NSU) . . . will be assisting on-
line. . . .

Dene Grigar
University of Texas at Dallas
```

All total, fifty people gathered for my online dissertation defense from places as far-flung as Australia and England, as well as around the United States. As the minutes ticked by and people logged in, I realized that I would be defending the last four years of my research, defending the start of my academic life, to all of them, many of whom I did not know and many of whom I could not *see*.

When Dene told me she planned to defend her dissertation in Lingua MOO, a virtual space written into reality by its inhabitants, I was very interested. After all, I had tried, unsuccessfully, to have my own dissertation defense conducted via a computer conference (or newsgroup) where my examiners and I would conduct our discussion of my research through a series of asynchronous messages posted back and forth. My own dissertation, "Talking around the Electronic Campfire: An Ethnography of Writing Teachers Investigating Computer-Assisted Composition within a Computer Conference," examined how participants in an online graduate-level English course interacted in an asynchronous electronic environment and developed a collaborative learning context to build a socially oriented and productive community. I thought my proposal for the format of the defense was appropriate to the nature of my study.

My chairperson discouraged it, however, saying that the department chair would not appreciate—*allow* was the intended meaning—such an unorthodox approach. As a result, my doctoral dissertation defense was successfully conducted around the traditional seminar table, face-to-face.

Generally, we regard a dissertation defense as an oral examination of a candidate's expertise before we confer professional status on her. It is, as we all know, a fundamental prerequisite of our profession. In a dissertation defense, a candidate presents her research to a committee of experts and other interested parties physically in a room, before fielding questions from this audience regarding her work. Once completed, the experts decide whether the candidate has achieved the level of thinking and writing necessary for being a productive member of the academy. The historical significance of the defense cannot be overstated.

John nods.
Dene holds up a sign.

According to R. S. Rait in his book, *Life in the Medieval University* during the Middle Ages "the Doctorate became an order of intellectual nobility with as distinct and definite a place in the hierarchical system of medieval Christendom as the Priesthood or the Knighthood." The series of private and public examinations the candidate had to pass guaranteed intellectual competence and his moral strength. . . .

The medieval "private examination" consisted of a "private exposition" given by the candidate to the doctor in charge of the candidate's education and, later, to a small committee of one of two other doctors from the university. Once the candidate passed this exam, he was ready for the "public examination" conducted in the cathedral and in front of a large audience comprised of colleagues and teachers from the university. (1931, 29-33)

We can see the private examination still present today in the series of meetings candidates have with their committee during the time they are working on their dissertations—"the dissertation hours" the academy

requires of students. When students complete their projects and are allowed to schedule their dissertation defense, there is an assumption that the handful of professors who have been guiding the candidate's work privately have given it provisionary acceptance. Ready for the public examination, the official dissertation defense, candidates defend their research in an open forum with colleagues and others who may wish to attend.

In the last few decades the public portion of the dissertation defense has moved increasingly toward very private affairs with only a handful of examiners posing questions to a doctoral candidate. Sometimes just the members of the committee who have already questioned the candidate in private meetings are the same ones examining her in the public forum. These meetings are generally conducted face-to-face and are not augmented or enhanced by the use of computer communications technology.

Some observers of my online dissertation defense viewed it as subversive because computers were used during my examination, while some saw it as a move back to a *more* traditional format than currently the fashion because it allowed for a public examination of a candidate.

Date: Thu, 03 Aug 1995
From: Katie Gilchrist
To: Dene Grigar
Subject: RE: The Defense

Dear Dene,
There was in interesting reaction from the [older] people [at Oxford University] I talked to on the conference - alot . . . thought that this was going back more to the old-fashioned sort of open viva, rather than the pattern now in this country of it being just you and two examiners in the room. It was the middle age range who were most dubious about the idea—those who have never really got to grips with computers.
Katie

As Dene says, the central components of academic life are research and the presentation and defense of new ideas. Scholars often devote years to the process of collecting, interpreting, and then presenting ideas and information either in written form as journal articles or books or in presentations at professional meetings. Thus, because of the time commitment, intellectual investment, and risk involved, as well as the fact that academic careers are often founded upon the successful presentation and defense of new and/or different interpretations of existing knowledge, presenting and defending one's research or interpretation can be viewed, and are often considered, literally as defending one's academic life.

Historically, dissertations and other investigations have been conducted primarily through public face-to-face hearings or discussions. With the advent of professional journals, authority for sanctifying and assessing academic endeavors became more print-based. Facilitated by such iterations of networked-computer communications as the Internet and the World Wide Web, research, presentation, publication, and defense of academic endeavors are moving increasingly into computer-created, defined, and augmented spaces like bulletin boards, electronic mail discussion lists, and MOOs—social virtual-reality systems that allow multiple and simultaneous users to communicate and interact in pseudophysical surroundings.

Let me make myself clear here. My dissertation defense was *not* subversive simply because I used computers. It undermined the tradition because of other choices I made. The medium I chose to use (Lingua MOO) and the way in which my collaborators and I decided to use this medium both contributed to the way my online dissertation defense undermined tradition. Before John and I discuss this issue, we should explain the reasons behind defending my work online.

Why Online?

Dene and John show "Why Online?" slide #1.

****

First and foremost, we wanted to demonstrate how vital computer technology is to scholarship, and at the same time forge new territory in electronic discourse.

****

Date: Thu, 13 Jul 1995
From: Bruce Jackson
To: classics@u.washington.edu
Subject: Re: Invitation to an On-Line Dissertation Defense

The usual Ph.D. oral is predicated on how much everyone in the room knows, how well the candidate can answer questions, and how adroitly the questioners can formulate their questions. You've added the ability to type well and quickly and you've subtracted body language and grunts and snickers and harrumphs of approval. It should be interesting to see how this plays out.

Bruce Jackson

My dissertation, "Penelopeia: Penelope in Homer's Story and Beyond," traces the way Homer's Penelope has been portrayed in literature, the visual arts, and musical composition from the Middle Ages to the present time. A large undertaking, it required me to investigate the arts from a six-hundred year period and be adept at scholarship for specific movements in its history. I needed to be familiar with literally hundreds of artists, many of whom were obscure or forgotten.

Locating this data presented problems in the early stages of my research. Using traditional print sources, I discovered little in the way of primary sources that focused on or referred to Penelope. However, when I turned to electronic resources, such FirstSearch, the World Wide Web, and virtual libraries, the number of references soared from forty-six to over three hundred. At the time I was working on this research, few scholars were actually using technology in this way. In fact, my web page, which offered updated lists of references to scholars, was one of the first of its kind available.

Because many of the works I located were obscure references with partial bibliographic information, it was imperative that I contact experts working in a particular historical period, art form, or artist. So I posted my web page as a work-in-progress and invited Homeric and humanities scholars from the various listservs I belonged to to look at and comment upon my research.

Although few scholars turned to the Internet for serious research into the Humanities in those early days of personal home pages, I did receive numerous responses. Many dates that were missing, artist names

that were incomplete, or artists so obscure I overlooked them were filled in by scholars visiting my site. Even after my dissertation defense, colleagues continued to send me references.

```
Date: Thu, 03 Aug 1995
From: Katie Gilchrist
To: Dene Grigar
Subject: RE: The Defense

Dear Dene,

This is rather late, but I noticed one omission from your
list of sources. Have you looked at Robert Graves' novel
"Homer's Daughter" (1955)? It's a rather odd rewriting of
the Odyssey, and shows up some of Graves' rather odd ideas,
but it is interesting—the narrator is Nausicaa, who
composes the Odyssey (following Samuel Butler's theory),
and claims to re-writre Penelope as a chaste heroine rather
than the wanton mother-goddess figure Graves thinks she
was originally. I also notice that you were still chasing
details on a book by Marion Chesney, called Penelope. She's
a British writer (Scottish, to be precise, I think),
female, and she tends to write historical romances—I would
expect that this book will be of that type (generally
regency period, but she might have a wider range). Useful
for when you produce the book, perhaps?.

Katie
```

Thus, because my work was always online, it seemed logical to present the defense of my research electronically.

This kind of investigation was relatively new at the time. Many of Dene's committee members had little training with electronic research or experience with the software programs and technology she used. It was important that they see the breadth of her understanding of innovative research techniques that led to a thorough investigation of her subject. The MOO defense, with its WWW interface, was important for making them aware of her scholarship and the potential that these resources held for their own work.

Dene and John show "Why Online?" slide #2.

* *

Just as importantly, we wanted to break new ground in the use of MOOs for research and presentation of knowledge.

* *

My understanding of Penelope and the works surroundings her were so profoundly shaped by technology that I wanted to continue exploring its effect upon academe. In particular, I was interested in how writing and thinking, activities underlying academic discourse, changed in electronic environments. Along these lines I began organizing a conference on Penelope for other scholars working in this area. Planning this online conference also influenced my decision to experiment with a MOO for my dissertation defense.

I was aware that Joel English, a graduate student at Ball State, had successfully defended his master's thesis online at DaedalusMOO. But as far as any of us could tell, my defense was the first one held online by a Ph.D. candidate. Although I would never discount the importance of English's innovative approach to his work, I do think that defending the terminal degree online holds broader implications for the profession because it is the last step in completing one's graduate work before embarking on an academic career. And as mentioned earlier, our profession places great stock in this event. Thus, any move outside of its conventions, many of which were carefully devised by medieval scholars so long ago, is breaking new ground, not just for the field but for the entire profession.

Date: Sun, 16 Jul 1995
From: Mick Doherty
To: Dene Grigar
Subject: Re: On-Line Defense

Wow, Dene! Good Luck!

I admit that I have no background whatsoever in your field of study, but I may show up and lurk just because the very

```
idea of an online defense is incredibly bold and
interesting.

Is an abstract of your diss somewhere on the WWW so people
who want to go aren't completely unprepared for the content
of the discussion (even if we're more interested in the
context, format and medium <grin>)?

Good Luck!
Mick D-
```

Dene's colleagues expressed interest in her work, her upcoming online dissertation defense, and the implications suggested by such an undertaking. As for myself, I was curious how Dene's approach may be used for my own scholarship, particularly in asynchronous electronic environments. I was also interested in learning more about how people may use MOOs for interaction and in seeing what could be done with a MOO as a site for the collaborative exchange and discussion of knowledge.

```
Dene and John show "Why Online?" slide #3.

    *********************************
    Defending this dissertation online also reflected
      an interdisciplinary background, which
      candidates at Dene's university must demonstrate
      in their work.
    *********************************
```

The School of Arts and Humanities at the University of Texas at Dallas is an interdisciplinary program, and its students are expected to develop a background and expertise in aesthetic studies, literary studies, and the history of ideas. Dissertations must reflect at least two of these fields, as must our dissertation committees. These requirements, and the fact that I had published articles dealing with both literary studies and computer technology, led me to choose a dissertation topic that not only included philosophy, history, Homeric studies, literature, visual arts, and translation theory, but also employed the use of the Internet, the World Wide Web, and MOOs—computer technology I used to facilitate my research.

From the email messages I received, it was obvious that my online dissertation defense interested scholars from many different fields and, in some cases, brought them together to discuss its usefulness at their own institutions.

```
Date: Fri, 14 Jul 1995
From: Mike Fraser [computer studies, Oxford University]
To: katie.gilchrist [classical studies, Oxford
University]
Subject: VivaMOO

Dear Katie,

A recent message on the Classics list mentioned that you
were intending to be present at Dene Grigar's thesis
defence. Unfortunately, I don't think I will be able to be
there but I would be most interested to hear how it
proceeds. Do you happen to know if anyone will be logging
the discussion, for example?

With thanks,
Mike
```

Dene and John show "Why Online?" slide #4.

* *

Lastly, Dene wanted to mark her territory and introduce herself to other scholars in the field.

* *

In attendance at my dissertation defense were numerous scholars working in the field of technology from universities, such as Massachusetts Institute of Technology and Rensselaer Polytechnic Institute, literary scholars from places like University of California at Berkeley, and classicists from Oxford University in England. People whose work I had long admired, such as Theodore Brunner and Randall Stewart, sent congratulatory messages after it was announced I had passed my defense. Shortly after my online dissertation defense was completed, I was invited to an online open house sponsored by the James Joyce Society to discuss

my findings on Joyce's view of Penelope. As a graduate student trying hard to establish myself in the field, I benefited greatly from this attention.

Some of the participants at my online dissertation defense had met me previously at conferences and workshops where I presented papers, but many others who attended had never met me or seen my work. They came because they were curious about my dissertation defense or were interested in my work on Penelope.

Date: Fri, 11 Aug 1995
From: Roni Keane
To: Dene Grigar
Subject: Great!

Dene,

BTW, is there a way I could have an unedited transcript of your online defense? I found the interactions intriguing. . . .

Congratulations! Hope you've had a chance to wind down a bit before school starts.

What Dene demonstrated was that making and sharing knowledge in a MOO environment can be a collaborative, social activity, simultaneously available to a great number of people. An online dissertation defense, such as Dene's, captures attention in a way that a traditional face-to-face defense cannot because it allows for more interested people to attend and participate via telepresence.

The Evaluation

Holding a dissertation defense online in a MOO is risky, not only for the candidate who must prepare and execute such an event, but for the profession as well. Taking defenses outside their conventional setting risks profoundly changing one of the most fundamental and traditional aspects of the university—that is, the way we examine its potential members.

Dene and John show "Evaluation" slide #1.

* *
First, this online dissertation defense was collaborative.
* *

Start: Tuesday, July 25,1995 4:02:50pm Lingua time (CDT)

[PANEL] Cynthia says, "Welcome to Dene Grigar's on-line dissertation defense"
[PANEL] Cynthia says, "Dene is ready to begin introductions and shortly she will prompt some slides to explain to our online audience what she is explaining to her committee . . ."
Kathy-Downey quietly enters.
[PANEL] Cynthia says, "thank you for coming and we will let you know when we may begin taking questions from the audience."

Dene shows slide #1 on MOO slideprojector.

* *
Penelopeia: The Making of Penelope in Homer's Story and Beyond
* *

Kathy-Downey finds a seat in Auditorium.

Dene shows slide 2 on MOO slideprojector.

* *
Thank you for joining us today for this unique event. . . .
* *

Cynthia Haynes acted as the online commentator for the event. She prompted the MOO slides containing the text of my oral presentation that I created for the online audience and explained what I was doing in the computer lab in front of the face-to-face audience—while also providing

technical assistance. Her workstation was not far from the main terminal where I was standing. But in order to facilitate the online portion of the dissertation defense, she had to move about the room. So her presence was highly visible and took the face-to-face audience's eyes off me from time to time.

It is not common for outsiders to have this kind of input into a dissertation defense, nor is it traditional for the candidate to have outside assistance when she is defending her dissertation. Dissertation defenses, by their very definition, imply that a candidate must struggle to find strategies to protect herself from attack by an outside force—in this case, a committee of expert who holds the beginning of her academic career in their hands. And she is expected to accomplish this *alone*.

Cynthia's collaboration with me extended beyond showing MOO slides and explaining what I was doing in real-life to my online audience. Along with the other creator of Lingua MOO, Jan Rune Holmevik, we worked for months organizing and designing the layout and presentation of my online dissertation defense.

```
Dene holds up a sign.
```

```
-- Start: Sunday, July 9, 1995

Jan says, "let me just recapitulate . . ."
Jan says, "This room can be formatted in open or
  panel format . . ."
Jan says, "in open format, like it is now, everyone
  can speak freely . . ."
Jan says, "in panel format people in the audence can
  only address the speakers via a moderator . . ."
Jan says, "the audiance will be able to talk amongst
  themselves but the speakers will not hear this,
  and this not be interrupted . . ."
Jan says, "I was thinking that during the defense
  you and your committee will sit up here in the
  panel, in panel format . . ."
Jan says, "so that no one can interrupt the
  meeting . . ."
Jan says, "but only listen . . ."
Cynthia [to Dene]: And I will be in the classroom
  helping your committee do all of this, Dene
```

This collaboration indicates the ability of MOOspace to extend the boundaries of time and space. Jan provided all of the online planning and technical organizing from his office at the Institute for Studies in Research in Higher Education in Oslo, Norway, while Cynthia and Dene worked from their offices in Richardson and Dallas. During this time, they never met face-to-face to discuss the dissertation defense. All planning took place online.

Jan's contribution also included designing a special room—the auditorium—located in the Conference Center at Lingua MOO (fig. 10).

As you know, "hearing" in a MOO means "seeing" text on a computer screen. So that I could concentrate on the questions asked me and avoid the confusion of the constant stream of scrolling text usually found in MOOs when large numbers of people gather and talk, Jan constructed this room with the option for a moderated forum. At the front of the room he created a stage where the panel could sit. Those of us sitting on the stage could not hear the conversations taking place in the audience. Audience members who wore special headsets, likewise, did not hear the conversations going on around them. They could only hear the panelists. However, when they took off the headsets, they could hear everything said. Panelist's questions were queued to Cynthia, who acted as moderator. From those in the queue, I would choose questions and the order in which I wanted to answer them.

In addition to this collaboration with Cynthia and Jan, I also worked with others on the day of the defense. John and Jeff Galin assisted the audiences with communicating in a MOO. John worked with the face-to-face audience in the lab in Dallas. From Pittsburgh Jeff helped the virtual audience.

Dene asked me to sit with her committee members and others in the computer lab at UTD and help them navigate and communicate within the auditorium at Lingua MOO during the online portion of her dissertation defense. Some of her panelists were not prepared to deal effectively and efficiently with the uniqueness of her online dissertation defense. I spent most of my time showing them how to make sense of the scrolling text that filled their computer screens.

Although MOOs and other forms of interactive computer technology will continue to develop and change, making it easier for people unfamiliar with programming to use, we may still require others to assist us in overseeing their operations or managing their use. Breakdowns,

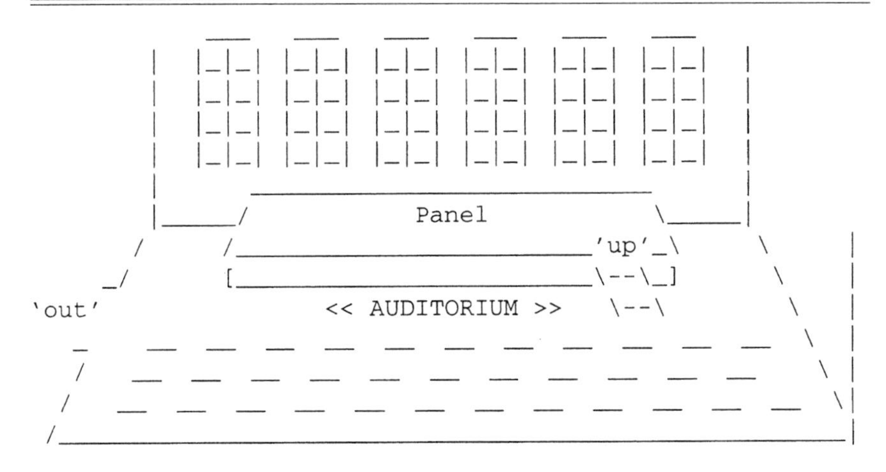

Fig. 10. Lingua MOO auditorium by Jan Rune Holmevik

minute-to-minute fine tuning, and troubleshooting are problems candidates do not have the time to deal with while they are in the middle of answering a question from a committee member. One of my reasons for seeking collaboration from so many colleagues was an attempt to save time and head off any potential malfunctions.

And a malfunction is precisely what *did* happen.

After I began my presentation, my collaborators discovered that the MOO slides I had prepared were not being received by some online participants. When Jan realized the MOO slides were not visible to everyone, he immediately began working to solve this problem, which was caused by a bug in the auditorium software triggered by the large number of people in the room and the fact that a slide projector was used to broadcast information.

> jeff says, "I assume Dene is doing her thing for the folks in UT at the moment."

> John says, "dene is explaining the background of her dissertation to the people gathered in the lab."

Standing at the podium delivering my presentation to the audience in the lab, I did not know that this was happening, nor could I have done anything about it if I wanted to. But we anticipated problems and planned for Jeff and John to step in where they were needed. The result of our good planning was that my collaborators acted as commentators and kept the audience informed.

Everyone was talking at once, MOOslides were being shown, and, watching it all scroll down the computer screen, it was easy to feel that everything was happening, as jazz pianist Theolonius Monk used to say, "All the time." But it was soon obvious that everything was not happening for everyone all the time, and for some people not at all. Guests in the auditorium began commenting about not being able to see the MOO slides Dene had prepared. Jeff's comment made me think large numbers of guests were not receiving the information Dene had prepared for them—information about the implications and value of her research. I jumped in with my comment to everyone.

```
Dene and John show "Evaluation" slide #2.

    ********************************
    Defending her dissertation online expanded Dene's
      work beyond traditional boundaries of university
      and committee.
    ********************************

[PANEL] Dene says, "are there any questions?"
[PANEL] Dene says, "pat, do you have a question for me?"
mikes blushes
[PANEL] patricia says, "yes, Dene. How do you position
  H.D.'s modernist, disenchanted Penelope . . ."
```

I invited Dr. Patricia Moyer, a retired faculty member from Exeter College and a visiting scholar at the University of North Carolina at Chapel Hill who specializes in female characters of Homer, to participate as a panelist because of her expertise in this area. Also invited to join in as a panelist was Katie Gilchrist from Oxford University, who was researching Penelope for her own thesis.

But Pat did more than examine my interpretation of literary artists, she also came to my defense when she saw another panelist did not agree with my methodology—that is, my decision to present my work chronologically. Trained in classical studies and well versed in the way Homeric female figures have been portrayed in subsequent periods of time, Pat, talking to us from North Carolina, stepped in to defend me at this moment in my online dissertation defense—something unheard of in a traditional one. Even other participants were aware of this breach in tradition.

```
[PANEL] patricia says, "I found your chronological
  generalisations quite valid & methodology right on."
mikes giggles
Keith smiles at Panelist
```

This kind of situation could not have happened in a traditional face-to-face dissertation defense. A member of the audience would not have been permitted to make such remarks in support of a candidate. And had they insisted on inserting such comments in the ongoing dialog, they probably would have been ignored. Protocol and hierarchy would have been observed.

One of the apparent benefits of MOO environments, however, is that they level the playing field. Potentially, everyone has the same opportunity to speak and be heard. This is good and bad. Good in the sense that there is the potential for multiple voices and perspectives. Bad in the sense that these multiple voices and perspectives are delivered essentially simultaneously and may drown each other out.

To deal with this problem, Dene and Jan implemented a moderated forum, where panel members, sitting on the stage in the auditorium, would not see or hear the constant stream of written dialogue and interaction shared between members of the audience. This allowed them to concentrate on the discussion surrounding Dene's examination. However, they were able to make comments or queue questions for Dene to answer as they saw fit. Thus, since Pat was sitting on the panel, she was able to interject her support.

It's interesting to note that Dene's panelists spoke only when asking formal questions. None communicated as freely and frequently as Pat.

The reason some panelists spoke little during my dissertation defense is that they viewed this part of the proceedings as a demonstration of the technology I used in my research rather than a real defense of my work. I will discuss this in greater detail later when I talk about the double defense that I had to give.

But going back to Pat's participation, we can see that MOOs used in this way extend the notion of examiner from a few committee members sitting in a classroom to a wider audience participating in the event from across the globe. In fact, the expansiveness of the medium makes it feasible for candidates to work productively with committee members from other institutions far from their degree-granting universities.

This concept of the extended classroom is not new. Scholars developing telecourses and distance-learning environments have been exploring this avenue for teaching. For example, at the annual meeting of the Association of Linguistic and Literary Computing and the Association for Computers and the Humanities held in Bergen, Norway, in 1996, Harald Ulland and Geir Pederson demonstrated an electronic classroom developed for distance learning. This project, a joint venture between the University of Bergen and the University of Oslo, makes professors and specific course work available to students at both universities. Their classrooms reach across hundreds of miles and offer students opportunities for learning that they would not have otherwise. MOOs too represent an alternative system for developing distance learning environments.

Dene and John show "Evaluation" slide #3.

****
This online dissertation defense required a new kind of preparation for me, as well as members of my committee.
****

[PANEL] Panelist says, "You quote Parker's Penelope and highlight the last line 'The will call him brave.' You say that it has an ironic tone and that the voice of Parker's Penelope exudes sarcasm and bitterness. You state further that Parker's treatment of Penelope is ironic. Further along in your dissertation you state that Parker's Penelope is savvy enough to recognize that history will not be kind to her, and that Parker's Penelope resents the way she will be remembered in literary history, although she is not angry over her husband's absence. Please define what you mean by irony and defend you assertions that Parker's treatment of Penelope is ironic. What evidence can you provide for your claim that Parker's Penelope is sarcastic, bitter, and resentful?"

One panelist wrote a lengthy, complex question usually found in oral defenses. But in a MOO environment, reading a long, scrolling stream of text presents problems for the audience. As most of us know, online writing demands we find new ways to present our ideas. Some people prac-

tice chunking—that is, writing in a series of short sentences in order to make text easier to read.

But some fundamental problems arise when we chunk text, especially text representing an in-depth response to an examination question. During my online dissertation defense I found that breaking up the information in this way made it difficult to concentrate and answer questions coherently. The process of typing a few words and an ellipsis, then hitting the send command, then repeating this process over and over again distracted me. Though I had spent much time in MOOs writing under pressure before, no time had been as important as this one, and never did I have quite the audience as I had on that day.

Dene's experience points out a very real problem we face when using computer-text mediated contexts for the dissemination and defense of large amounts of information: The lack of familiar notions of navigation. With a book or journal article, it is easy to quickly gauge its length. With electronic text that spans more than one screen display, it's possible to see the whole text only in one's mind, a feat made more difficult by the necessity to keep track of multiple, individual screens of information. Working with print-based information, we can easily flip back several pages to, for example, the bottom third of a left-hand page where we remember reading a salient point. In a computer-mediated context, the text is essentially a long scroll moving up and down behind the screen. Unless the text is augmented with hypertext links allowing us to jump forward and back through the text, we are forced to scroll through a long, linear banner of text, looking for the information we want or need, whether assisted by our own speed-reading abilities or the search capabilities of the software we are using to display and manipulate the text.

Furthermore, there are problems associated with reading and decoding large blocks of text displayed on a computer screen where the total number of lines of text can be limited by windows and menu bars or buttons, or worse, the insertion of someone else's comments.

Dene's chunking strategy demonstrates a technique of composing our comments in smaller, manageable blocks that can be quickly displayed and digested on computer screens and, arguably, replicates a technique we use in oral presentations: speaking in short phrases . . . and using words or paralinguistic features . . . as, um . . . placeholders . . . while we simultaneously formulate our thoughts.

Dene and John show "Evaluation" slide #4.

* *

By reaching beyond the university, this online dissertation defense served as Dene's formal introduction to the field at large.

* *

As I mentioned earlier, one of the reasons for holding my dissertation defense online was to introduce my work to other scholars, to mark my territory. In this sense my online dissertation defense, held in Lingua MOO's auditorium with fifty people in the audience, resembled a conference panel with a new scholar (me) presenting her first work to colleagues (my audience) in her field. And like conference proceedings, the online dissertation defense makes research available immediately, often long before the dissertation is conventionally archived.

All the scholars collaborating with me on this event were exploring, in similar ways, the potential of MOOs for sharing knowledge with a large body of people. As was pointed out earlier, John's dissertation focused on this very topic. Jeff developed MOOCentral, a resource guide for MOOs, and has conducted considerable research on this topic. Cynthia and Jan, as the creators and managers of Lingua MOO, have long participated in and organized meetings held in MOOs. For example, C-FEST, a series of online gatherings that they instigated at Lingua MOO, resembled round-table discussions much like those we find at traditional academic conferences.

Dene and John show "Evaluation" slide #5.

* *

Archived in LinguaMOO, this online dissertation defense provides an historical account of the event.

* *

[PANEL] Dene says, "the fascination with Penelope began after the Middle Ages, in the Renaissance when printing became more prevalent. . . "

[PANEL] Dene says, "artists seem to be interested in her ability to persevere for 20 years. . ."

Richie takes off his headset. He will now be able to hear everything that's being said in Auditorium

[PANEL] Dene says, "that she could possibly remain faithful for the time her husband had been away . . ."
[PANEL] Dene says, "that she could have so many men interested in marrying her . . ."
[PANEL] Dene says, "that she could deter so many men from taking over the palace, killing her son . . ."
[PANEL] Dene says, "that she could outwit Odysseus in the end . . ."
[PANEL] Dene says, "that is the man whom Athene called the most cunning of men . . ."
[PANEL] Dene says, "this is why they are interested in her."

We made a record of the ideas Dene discussed during her online dissertation defense, as well as comments and questions posed by her audience. Archived in an electronic repository, others can now study and utilize it for their own work. Of course, taping Dene's performance was not without its drawbacks. Knowing that what we were saying and doing was being retained for posterity put great pressure on all participants to show themselves well to the audience viewing the event.

In light of this, it was imperative that my examiners come to my dissertation defense prepared with well-thought out questions and for me to devise strategies for answering questions far in advance of the event—an event that included writing my answers to an audience.

I found the concept of writing the dissertation defense disconcerting for two reasons. First, writing one's presentation and answers in a dissertation defense runs counter to the notion of an oral examination. Second, writing in a MOO is not like writing in real life. It is vastly different.

Dene and John hold up a sign.

As Jan and Cynthia point out in their paper, "Synchroni/CITY: On-line Collaboration, Research and Teaching in MOOSpace," presented at the 1996 Modern Language Association meeting:

Jan says, "One of the major advantages of writing on the MOO is that you are literally and actually IN the text while writing it. . . ."
Cynthia says, "The textual nature of the environment enhances and actually evokes writing, so that writing IS the landscape and medium of communication among people at Lingua. As Jan said, the capability of our text to be written online, stored online, and revised online means that the same activities are possible for teaching, for recording meetings, performative events, and for editing those transcripts and archiving the final copies, not to mention the ability of those texts to be played back online and/or emailed to participants."

Writing as landscape—a strange concept and a stranger experience. I was standing at the computer terminal in the lab in front of my real-life audience, writing my answer to a person sitting in the auditorium of Lingua MOO, who, like me, was represented by text. We were essentially text writing text. And we, as both personae and text, would be stored online, revised online, played back online, and emailed to others online.

As Dene suggests, MOOs constitute virtual existences written into reality by participants interacting with computer technology. They are architectural and, as Cynthia and Jan have posited in their research on MOOs, "architextual," allowing for a dynamic facsimile of human intellect and personality to be created, preserved, and interacted with within a computer system.

Dene and John show "Evaluation" slide #6.

* *

Because this online dissertation defense introduced Dene to a large body of scholars and represented her preliminary thinking about a topic, it emphasized UTD's responsibility to produce a well-trained professional.

* *

As I defended my dissertation online, I was aware that the whole university was defending with me. Though this may be true to a certain extent for traditional defenses, it was never more apparent than when my thoughts and ideas were being read by a host of other scholars from across the globe. At that point I represented the shaping and teaching of members of my committee. That I was the product of their handiwork put a great deal of responsibility on my committee—that is, before I even had the opportunity to grow intellectually in my field, to grow comfortable with my work. As I fumbled one of my panelist's questions because the question, one long stream of unbroken text, was difficult to read as it scrolled across the screen, I was seized with the fear that if I didn't do a good job, I would embarrass my teachers and myself in front of potential colleagues.

Looked at from a more economic perspective, online dissertation defenses like Dene's are dramatic demonstrations of profit (favorable recognition for candidates and examiners, and by extension their universities) derived from the sale of marketable goods (the production of individuals who will promote the orientation and mission of their degree granting institutions) after the investment of capital (physical and intellectual resources controlled by the university and its faculty). As in any market, academic goods perceived to fall below the accepted quality standards may fall from demand, and their producers may fall out of favor with consumers. In a job market that may well continue to be grim, it behooves candidates, faculty, and administrators, considering the use of MOOspaces for the public presentation and defense of their work, to think carefully about the ramifications. Well-planned and executed public defenses of academic work may attach a cachet of higher quality to individuals and institutions associated with these events. Problematic and/or ill-prepared online defenses may cast negative values on those associated with them.

Dene and John show "Evaluation" slide #7.

```
*********************************
This online dissertation defense required that
  participants think in advance about the way they
  were going to present themselves online.
*********************************
```

Beyond how we ask questions, we also must be concerned with how we present ourselves online. Comments we would make in a small group or with close colleagues who understand the context we are working in are not necessarily appropriate to a larger unseen virtual audience unfamiliar with us or our presentation style. Voice intonations, body language, and facial expressions can be lost in a MOO environment unless the sender of the message types those kinds of messages. Without them, we can misinterpret the spirit of the message.

```
[PANEL] Panelist says, "Dene, it seems to me that you
  have approached this material primarily from close
  reading of texts, both visual and verbal, with only
  fairly general consideration of the contexts of the
  works. I wonder whether, especially in the case of
  works from the earlier periods, when attitudes might
  be significantly different from today's, you would
  consider it important to consider in greater depth
  the historical contexts. I'm thinking, for example,
  of your assumptions that the Renaissance Penelope
  represents a more liberated reading of the character,
  this at a time when most contemporary historians
  believe that attitudes towards the position of women
  in society had changed little from those of the
  medieval period, or had even become more
  conservative."
jeff says, "whew what a question. Seems like the moo
  makes folks go right to the point"
mikes [to jeff]: i've seen some diss roasts.
RL can be nasty
```

Questioning a candidate's stance on an issue is part of any dissertation defense, and it can indeed be "nasty." Therefore, it is imperative that we think through our comments and questions in advance so that when we do question a candidate, we come across to our audience as professionally as possible.

Candidates and committee members must be aware that the strategies used in a traditional dissertation defense differ widely when they move into a computer-augmented environment. The major difference is that in a traditional dissertation defense the audience is comprised of the committee members and a few close colleagues. Personal conflicts among committee members or ambivalence of a committee member toward a candidate's work does not generally extend beyond the physical walls of the examination room.

```
Date: Fri, 14 Jul 1995
From: Patrick M
To: classics@u.washington.edu
Subject: Re: Invitation to an On-Line Dissertation Defense

I thought the real trick was to get Prof. X and Y, who hate
each other's guts, to spend the whole time debating each
other while the candidate escapes unscathed.

(Just kidding!) In truth, this sounds like an interesting
approach.

Patrick M. Thomas
Dept. of Archaeology/Art History
Univ. of Evansville
```

Even in the best situations, and despite the best preparations, it is easy to look foolish online, to miss a cue. Jan, Cynthia, and I worked very hard to organize the environment of my online dissertation defense so that I would not be interrupted while I was answering a question. As I said previously, panelists were told to queue a question to me using special commands. But one question was queued before I had finished answering another.

```
[Panel] Dene says, "this was something that did not
  appear previously in the Middle Ages. . ."
[Panel] Dene says, "Although i am not suggesting that
  the Renaissance period is one of great sexual
  freedom, I do believe that the issue of freedom does
  indeed become . . .
[Panel] Dene says, "important. . ."
[Panel] Dene says, "freedom in regards to what is means
  to be human . . ."
[Panel] Dene says, "And the modern period, the idea of
  intellectual freedom in regards to Joyce, Pound. . . ."
```

The moderator says, Question from a Panelist: You have given us a lot of historical information in your dissertation for each period that you deal with. What do you think is the conceptual frame of your dissertation that would distinguish your dissertation from a history of Penelope.

[PANEL] patricia says, "I found your chronological generalisations quite valid & methodology right on . . ."

[PANEL] Dene says, "I have tried to approach this study as an exploration . . ."

[PANEL] Dene says, "I came to the work with no preconceived notions of what I would find . . ."

[PANEL] Dene says, "perhaps what sets it apart from a historical approach is this idea of exploratory writing . . ."

jeff says, "come-on Dene go with the question . . ."

Jeff was correct in admonishing me to "go with the question," and had I seen his comment during the defense, I would have surely become anxious about the way I was being perceived by my colleagues and the way the defense was going. Faced with an additional question before I had completed the previous one I was caught off guard, and it looked to my audience that I was hedging or unable to give an appropriate response. In sum, I, along with the panelist who had asked the misqueued question, looked a bit foolish.

Writing on our feet runs counter to the idea of good scholarship, which most of us see as a careful crafting of written work. There is nothing careful about having to respond in a split second to a committee and a universe full of strangers who know very little about our work—and to do so quickly, brilliantly, and error free.

Dene and John show "Evaluation" slide #8.

* *

This online dissertation defense required special training for committee members and other participants so that they could perform effectively and efficiently in a MOO environment.

* *

Weeks prior to my dissertation defense, Jan and I hosted several training sessions for Pat, Katie, and members of my committee. We met online—Pat from her office in North Carolina and Katie from hers in England—and practiced conversing with one another and queuing questions. Pat and Katie attended all training sessions I offered. They even visited LinguaMOO on their own to get better acquainted with the environment. So that others could participate more fully during the event, I also trained members of my listservs for the event. Jan and I both prepared manuals for those who needed directions for communicating. People who were nervous about their own performances as audience members met at Lingua MOO and practiced for my dissertation defense. In total, I spent approximately fifty hours training others to use Lingua MOO in preparation for my defense.

But candidates generally do not worry about training others in a traditional dissertation defense. In fact, it is usually the *candidate* who must be coached by her committee chair in preparation for the event. In my case both were true: My chair advised me about how to defend successfully *and* I showed him, and others, how to participate effectively in a MOO.

Participants unfamiliar with a MOO environment were also unfamiliar with the excitement that this new environment brings. Like any newcomer introduced to a MOO, many participants engaged in personal discussions and fidgeted with the objects they could manipulate rather than focusing on the examination.

To be honest, the conduct of Dene's audience did not differ so radically from the old pros who meet regularly in a MOO. Anyone who has ever attended a C-FEST or a Tuesday Cafe knows that we *must* spend a certain amount of time making personal connections to mask the sterility of the environment. We emote and engage in informal talk because it is necessary for establishing the camaraderie we need in order to work productively together.

```
We see Dene writing something.
John holds up a sign.
```

```
Start: Monday, April 24, 1995 7:02 pm Lingua time
  (CDT)

Cynthia drops Machine (recording).
Cynthia says, "Hi . . . Welcome to Lingua MOO!"
```

```
Cynthia says, "Most of you know me . . . but this is
  Jan Holmevik, my partner in founding this
  MOO . . ."
Cynthia [to Jan]: Any wise words, Jan :)?
dene teleports in.
Cynthia waves at dene.
Walter sits down at South Table . . .
fred teleports in.
Jan says, "Well I just want to say that we are really
  delighted to have you all here today. . ."
Brice suddenly appears near you.

                                        _______________
                                       |               |
R.U.Rhetoricus? holds up a BIG sign:   | Hi Everyone!  |
                                       |_______________|
Cynthia waves at Brice.Brice waves at
  Lingua(helper), Jan, Bitstream(helper),
  Cynthia, Collin, Eric, Gilles [affable],
  R.U.Rhetoricus?, cath, Diane [wondie], Jim,
  David, Walter, dene, and fred.
Cynthia smiles.
Brice says, "howdy all"
Jan says, "and I hope we can have a fruitful
  discussion :)"
Cynthia [to Jan]: Thanks :) . . . let me get things
  started with a few short slides
Cynthia says, "before the introductions . . ."
```

People waving and smiling and holding up signs are essential communication for creating a sense of community among participants in a MOO session. However, this informality, which serves us so well in collaborating, meeting with one another, and hosting class discussions and student conferences, can work against us in a formal ritual like a dissertation defense.

Most every teacher who has introduced her students to the synchronous-communication capabilities of MOOs laments the amount of time spent seemingly off task, sending and receiving private messages.

Another lament is that the teacher seems hard pressed to compete for attention when students are staring intently into the spaces behind their computer screens.

The informality of MOOs may be derived from their origins. MUDs, a forerunner of MOOs, were developed in 1979 so that Dungeons and Dragons, a role-playing game, could be played by computer. Although MOOs are now being used for a host of more serious endeavors, they have lost none of their playfulness. In fact, it is precisely this balance of play and work that makes writing in a MOO exciting for both scholars and students.

Scholars have been exploring MOOs for teaching and researching for many years. Many of our colleagues helped to blaze the trail for others to use MOOs as serious spaces for their work. Those familiar with this environment understand the "pseudo-physical surroundings"—to borrow from John's earlier remark—that they are working in and are not concerned with the conventions that normally apply in real life. In fact, it has been necessary to adopt new conventions to this environment, and these are being developed as we see they are needed.

For example, in a MOO, conversation is immediate and fingers fly across keyboards to keep the talk alive. We write quickly to create a sense of reality and we don't care about good spelling, good grammar, or good typing. This environment is informal—the writing, fun.

```
[PANEL] Dene says, "Beccafumi, for instance, was
  working outside the 'Renaissance' tradition at the
  tinme he was painting Penelope. Although very little
  scholarship exists about his view of Penelope, many
  scholars discussed his emotional approach to art."
[PANEL] Katie says, "helo"
ChrisB says, "She "
ChrisB says, "OOps, Dene could be pulled two ways, eh?
  By new critics and new historicism . . . ?"
Jan says, "last speaker is here now"
```

It looks like I cannot spell or type the word *time* correctly, that Katie cannot spell or type *hello* correctly, and that Christine Boese can't complete a sentence. But, of course, all three of us *can* spell and type well, as well as complete our thoughts on paper. We know this about ourselves, and we know that others know this about us. But this informality seems strange in a traditional dissertation defense, which is a more formal

affair. So, moving it into MOOspace where we *write informally* poses interesting problems. Keyboarding, grammar, and spelling can haunt our performance.

This may be true. Moving the dissertation defense from the physical seminar room to the virtual auditorium may well present us with problematic situations. As is being shown in other online endeavors, interaction in cyberspace can be disconcerting. Sure, interaction in MOO environments can be hampered by transmission lags and echoes, as well as breakdowns that prevent any communication, but this will change for the better in the near future. Several computer technology companies released new software that allows us to conduct telephone calls over the Internet, as well as conduct real-time video conferencing. Keyboarding, grammar, and spelling may soon not haunt us anymore.

Problems Online

John and I have outlined several issues my collaborators and I had to face in the preparation of my online dissertation defense. Technical difficulties, distractions by participants, training for panelists and participants, and the informality of the environment all presented challenges for us. I should mention another area that posed the biggest difficulty for me, personally—that is, the double dissertation defense.

When I first approached my committee chair about using a MOO as the site of my dissertation defense, he readily agreed to it. He knew I had developed an expertise in using technology for research. And since he had written numerous letters of recommendation for me to teach composition, which at UTD was taught in a computer-augmented environment, he knew I had spent a great deal of time writing online and teaching others how to do so. When I suggested that we involve outside experts—that is, Pat and Katie—in the event, he also agreed. Though eminent scholars all of them, my committee's backgrounds ran the gamut of medieval literature, modernist poetry, translation studies, Renaissance art history, and classical studies. However, none were specifically researching Homer's Penelope. So, it made sense to bring in my two colleagues. He also was aware that Cynthia and Jan would need to work with me in order to organize the technology. I also told him that I would have other collaborators on hand to help participants in the lab and online.

We decided that I would begin my dissertation defense online, where I would introduce participants and guests to the research and methodology used to collect my data. After my presentation, we agreed that I

would open the floor to questions. Each committee member, as well as Pat and Katie, could examine me. Additional questions from other participants would follow. We also agreed that during the latter portion of my dissertation defense, members of my committee would ask me questions, and I would respond orally, face-to-face.

What I did not realize was that he expected the online portion of the dissertation defense to stop during the oral examination. I had fully planned for John and others in the lab to report to the online audience about what was taking place. I also expected to be informed when the oral portion of the defense would begin.

[PANEL] Dene says, "actually, artists like John Skelton recognized her 'wit' . . ."
[PANEL] Dene says, "though he may not have been aware of her outwitting Odysseus, he was aware that she had outwitted the suitors."
Sky finds a seat in Auditorium.
[PANEL] Cynthia says, "We want to conclude the online portion of Dene's defense now and thank all of you for coming . . . and apologize for the technical difficulties in the beginning . . ."
Kim gets up from her seat.
Kim tiptoes out.
ChrisB claps for Dene.
[PANEL] Cynthia says, "Please give your applause if you will to our panel :)"

To my audience, it seemed like the event suddenly ended. They were essentially left in the dark. Many of them stayed online to find out the results of the examination. Others sent email messages later inquiring about what happened or commenting about the length of the event.

Date: Thu, 27 Jul 1995
From: Jeffrey R Galin
To: Dene Grigar
Subject: Re: Dr. Grigar I presume . . .

Dene,
First of all, congrats. Another dissertation defended means there is hope for me too.

To tell you the truth, the online defense went pretty much as I expected, except it was shorter than I anticipated.

Even though I had prepared to answer questions both online *and* orally, I still had a difficult time making the transition from one environment to another. Articulating an idea in written form in a MOO and, then, suddenly having to respond orally to another question—essentially switching from virtual to real reality—took some time to get used to. And of course, in a dissertation defense, there is precious little time for a candidate to think, much less orient herself to a new environment. We are expected to know the answers, to perform smoothly, thus proving that we are qualified to receive our degrees.

Of course, any number of things can go wrong in a traditional, face-to-face dissertation defense as well. But, since there is more nonverbal and paralinguistic communication feedback, it may be easier for a candidate to adjust to the rapidly changing connotations associated with multiple questions. The fact that Dene felt disconcerted switching back and forth answering online and face-to-face questions supports the notion that computer-mediated spaces like MOOs promote essentially different communication and interaction contexts and thus demand further critical yet creative investigation.

Thoughts and Recommendations

Dene and John hold up a sign.

Many of the problems outlined above can be solved by just a bit of forethought and preparation. For others planning to defend online, we recommend that you do the following:

Dene and John show "Thoughts and Recommendations" slide #1.

```
* * * * * * * * * * * * * * * * * * * * * * * * * * * * * * * * *
Anticipate technical problems and make an
  alternative plan (or plans), and request ample
  technical assistance before and during the event.
* * * * * * * * * * * * * * * * * * * * * * * * * * * * * * * * *
```

We prepared for everything, from the server crashing to interruptions from scrolling text. Thanks to Billy Barron from our Academic Computing Center, we had a backup server on hand in case the one Lingua MOO runs on failed. And due to Jan's programming wizardry, I had a moderated forum to shield me from most distractions while I wrote my answers. But despite all of our foresight, we still encountered problems. As I mentioned, the MOO slides I had prepared could not be seen by all participants. Having John and Jeff on hand did alleviate this problem, though it did not solve it.

```
Dene and John show "Thoughts and Recommendations"
  slide #2.
```

```
* * * * * * * * * * * * * * * * * * * * * * * * * * * * * * * * *
Educate committee members about the MOO
  environment, especially if they have no
  experience with it, and encourage them to prepare
  MOO slides of their questions in advance.
* * * * * * * * * * * * * * * * * * * * * * * * * * * * * * * * *
```

While it is imperative that candidates defending online have their committee's support, it is even more crucial that they have the committee's enthusiasm. The commitment to training and a desire to understand the medium plays a big part in the way examiners respond to an online dissertation defense. Candidates need to know that their committee is communicating effectively in a MOO.

It would have been very helpful not only to me but to the members of my committee if they had created MOO slides of their questions in advance of the online dissertation defense. MOO slides would have alleviated their need to compose under pressure, would have allowed them to check their work, and would have been waiting in the queue long before the dissertation defense began. One member did create them; all others typed their questions to me during the event.

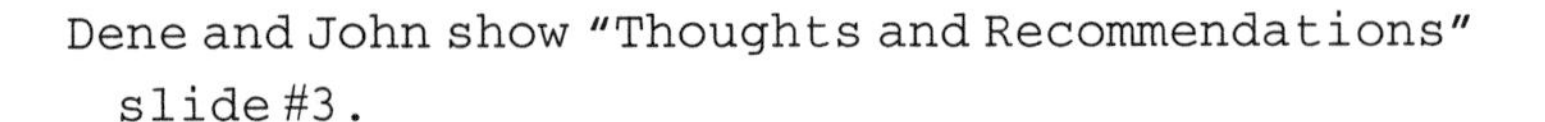

Dene and John show "Thoughts and Recommendations" slide #3.

* *

Be familiar and comfortable communicating in a MOO long before defending a dissertation online.

* *

It is also important to feel comfortable with the computer technology you will be utilizing, and depending on, for an online dissertation defense. In "Galin's Rules of Thumb," Jeff makes numerous recommendations about how and when to use computer technology in teaching and research, emphasizing that teachers be familiar with each application they expect their students to use.

Dene and John show "Thoughts and Recommendations" slide #4.

* *

Prepare to defend twice—once online and once in the traditional style, face-to-face.

* *

Candidates should anticipate that their committee may want to hear them expound upon an idea aurally. Dissertation defenses are a kind of test run to see if a candidate can stand up to public scrutiny in a face-to-face setting like an academic conference. Candidates should practice, then, moving from writing online to answering questions orally in order to prepare for a double defense, like the one I experienced.

Dene and John show "Thoughts and Recommendations" slide #5.

* *

Take advantage of interactive, collaborative, and powerful opportunities.

* *

The use of computer technology does not mean that the humanistic component of research, publishing, collaboration, or learning will suffer, but rather that these endeavors can be broadened. Conducting research and collaboration in MOOspaces can promote new connections and interactions not only between candidates and their examiners, but also between what these candidates are expected to become—teachers—and their students.

```
Date: Mon, 31 Jul 1995
From: Linda Wright
To: classics@u.washington.edu
Subject: Re: On-Line Defense

I hope Dene, who was defending her dissertation, will add
some comments to this brief summary of what I observed. At
first, yes, there was a bit of confusion and I thought one
person remained somewhat disruptive throughout. It could
be that he was not aware everyone in the "room" was seeing
his comments. Lingua MOO uses more features of the MOO
utility than I've seen before, and I never did figure out
how to put on my headphones.

There were very clear directions on how to get to the
correct room once you logged in, and then a "handout" by
your "seat" gave you some information on Dene's topic.
Everything seemed to run quite smoothly and just like a
real live defense. "Conversation" in a MOO is limited to
how quickly one can type, and such written dialogue does
differ from spoken exchanges, but a wider audience can
participate and "different" doesn't equate to "inferior."

I was satisfied and hope to see more of this in the future.
By the time I can defend, the technology should be in place
to allow me to have an inter-university committee.
Get ready. :)

Linda Wright
University of Washington
```

Despite the "confusion," the "disruptive" audience member, difficulty in figuring out the commands, the necessity to type well and fast, and the differences in "written exchanges," as well as the other

problems John and I have discussed, scholars still express the desire to defend their work online. We do not have to look far for the reasons why—they wish to expand their ideas and resources outside of space, time, and place, limitations imposed upon the discussion of their work. Like my collaborators and me, they recognize the potential interactive computer technology, like a MOO, holds for their future, their academic careers.

Conclusion

Welcome to Lingua MOO Auditorium.
You have arrived.

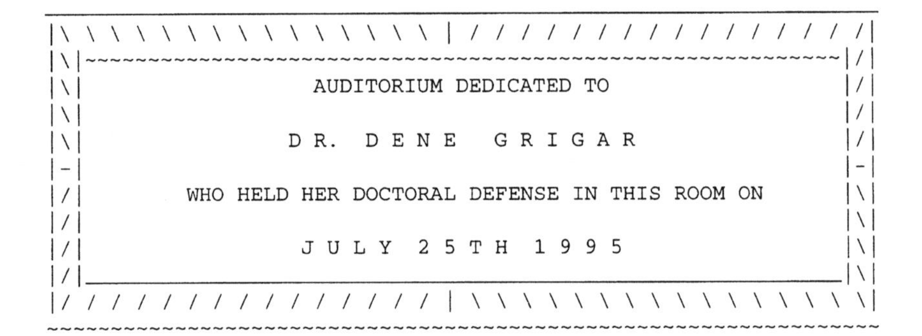

AUDITORIUM DEDICATED TO
D R. D E N E G R I G A R
WHO HELD HER DOCTORAL DEFENSE IN THIS ROOM ON
J U L Y 2 5 T H 1 9 9 5

You see a bronze plaque with black lettering mounted to the wall above the doors.

You see Dene and John standing on the stage at the podium. Papers are scattered all over the floor around their feet. They reach for one last page to read.

They say, "As we can see, the collaborative, social constructive, synchronous, and interactive nature of MOO environments can be unsettling in that it contests familiar paper with unfamiliar electronic text, verbal with visual orientation, oral with print interaction, and face-to-face with virtual context. Despite the problems we have outlined, there can be, arguably, a level of successful adoption of this medium for teaching and learning, as well as presenting and defending one's academic life."

They say, "But, new contexts and new paradigms take time to accept and adapt to personal productivity. Rather than providing complete answers as to how to utilize MOOspaces effectively, the implications and recommendations outlined here should be taken as invitations for further critical and creative thinking. We should also continue to examine and experiment with ways these computer-mediated spaces can augment our goal of making and sharing knowledge."

They say, "One way, as we have shown you, is the expansion of the notion of the academy beyond its physical reality and its limited opportunities for immediate and far-reaching collaborative exchange of ideas, beyond its traditional ivy-covered walls, and out to an electronic edge that can facilitate the inclusion of more voices in the sharing and discussion of ideas."

Dene and John hold up a sign.

Here is a list of collaborators who shared their ideas and knowledge via email messages, web sites, MOOspaces, and printed text.

"AEE Seminar." Telnet lingua.utdallas.edu 8888 (24 April 1995).

Barber, John. http://www.nsula.edu/jfbarber/diss.html.

Doherty, Mick. "Re: On-Line Dissertation Defense." Email to Dene Grigar (16 July 1995).

English, Joel. "Online Masters Thesis Defense." Online Master Thesis Defense Logs. Telnet MOOdaedalus.com 8888 (nd).

Fraser, Mike. "VivaMOO." Email to Dene Grigar (14 July 1995).

Galin, Jeff. "Dr. Grigar, I Presume." Email to Dene Grigar (27 July 1995).

———. "Galin's Rules of Thumb." http://lingua.utdallas.edu/moocentral/.

Gilchrist, Katie. "Re: The Defense." Email to Dene Grigar (3 Aug. 1995).

———. "Penelope Abstract." Email to Dene Grigar (17 July 1995).

Grigar, Dene. "Invitation to On-Line Dissertation Defense." mbu-l@unicorn.acs.ttu.edu (16 July 1995).

———. "On-Line Dissertation Defense." Online Dissertation Defense Logs. Telnet lingua.utdallas.edu 8888 (27 July 1995).

———. "On-Line Dissertation Defense and WWW Site." Personal email (19 July 1995).

———. "Penelope: Penelope in Homer's Story and Beyond." Available: http://www.utdallas.edu/~aca102/penelope.html.

———. "Re: Invitation to an On-Line Dissertation Defense." Personal email (13 July 1995).

Haynes, Cynthia, and Jan Rune Holmevik. "Synchroni/City: Online Collaboration, Research, and Teaching in MOOspace." Paper presented at the Modern Language Association Convention, Chicago, December 1995.

———. "Synchroni/City: Online Collaboration, Research, and Teaching in MOOspace." Available: http://lingua.utdallas.edu/~archive.html.

Holmevik, Jan Rune, Cynthia Haynes, and Dene Grigar. "On-Line Dissertation Defense Planning Meeting." Telnet lingua.utdallas.edu 8888 (9 July 1995).

Jackson, Bruce. "Re: Invitation to On-Line Dissertation Defense." Email to Dene Grigar (13 July 1995).

Keane, Roni. "Great." Email to Dene Grigar (11 Aug. 1995).

Rait, R. S. *Life in the Medieval University.* Cambridge: Cambridge University Press, 1931.

Thomas, Patrick. "Re: Invitation to an On-Line Dissertation Defense." Email to Dene Grigar (14 July 1995).

Ulland, Harald, and Geir Pedersen. "The Use of Distributed Electronic Classrooms in the Teaching of Language and Literature." Paper presented at the Joint International Conference of the Association for

Literary and Linguistics Computing and the Association for Computers and the Humanities, Bergen, Norway, June 1996.

——. "The Use of Distributed Electronic Classrooms in the Teaching of Language and Literature." Available: http://www.hd.uib.no/allc-ach96.html.

Wright, Linda. "Re: Invitation to an On-Line Dissertation Defense." Email to Dene Grigar (31 July 1995).

10

The Play's the Thing

Theatricality and the MOO Environment

Juli Burk

Introduction

When Shakespeare wrote, "All the world's a stage / And all the men and women merely players," he could hardly have known the stages human beings might inhabit at the end of the twentieth century. Close to four hundred years after Shakespeare wrote those lines, the stages on which men and women play are no longer limited to physical platforms in large rooms or open-air spaces. The advent of radio, film, and television brought stages for mass transmission of human players in action, and the development of computers, modems, dial-up connections, and ARPAnet, the Internet's ancestor, added cyberstages as arenas on which women and men might play. Today, cyberstages for human interaction on the Internet are plentiful, from asynchronous email, listservs, and bulletin boards to IRC chat rooms, to MUDs such as D&D (Dungeons and Dragons), Diku and other online games, to social MOOs, and to professional and educational MOOs. But, to transform Gertrude Stein's famous comment on roses, a stage is a stage is a stage is a stage. And in the Western world, any time the stage metaphor is employed, the vast majority of roads lead not to Rome but instead to Athens and the Aristotelian paradigm of dramatic action.

The educational MOOs that are the subject of this book have a lineage that goes back to the first interactive disk-based games and early online environments of the 1960s and 1970s, parents that share, among other things, an action-based narrative structure. And, interestingly, the first entry in Lauren Burka's carefully documented chronology, *The MUDline,* is for J. R. R. Tolkien's novel *Lord of the Rings,* published in 1937 (Burka 1995). As a feminist scholar, professor, theater director, and MOO archwizard, I find the patriarchal power relations of the narrative quest structure and the Aristotlean underpinnings of any stage metaphor not only linked but also disheartening. Aristotle's *Poetics,* written nearly two thousand years before Shakespeare's declaration about men and women on the world's stage, defines what has come to be known as theater in the Western world through his exegesis of the dramatic form called tragedy. At the same time, the *Poetics* articulates a patriarchal artistic tradition made famous by the Protagorean view that "man is the measure of all things," a measure against which women always and everywhere, no matter how talented, accomplished, or revered, fall short. In discussing the four most important aspects of character, Aristotle makes it clear that women are not as suitable for tragedy because they are inferior beings in whom manly valor is inappropriate (26). Defining tragedy in chapter 6 of the *Poetics* as "an imitation of a noble and complete action, having the proper magnitude" (11), Aristotle goes on in chapter 7 to define plot, the "soul of Tragedy" (13), as complete actions having a beginning, middle, and an end, placing linear narrative structure at the core of tragic action. Throughout the *Poetics* Aristotle refers to Sophocles' *Oedipus Rex* as an exemplary model of dramatic writing. In the play, Oedipus' kingdom faces a plague that the oracles he consults say will end once the murderer of the previous King is found and punished. Having solved the riddle of the Sphinx to gain the crown, Oedipus vows to solve this mystery only to discover that he himself is the guilty party. For the Western theater and particularly for Sigmund Freud, this play remains among the most powerful of quest narratives.

Teresa de Lauretis, in *Alice Doesn't,* has written persuasively about the impact of narrative on the female subject. Briefly, de Lauretis claims that narrativity is no more than an expression of Oedipal desire as outlined by Freud. As such, it requires a hero involved in a quest that demands an obstacle inevitably identified as feminine. The argument starts with Laura Mulvey's famous assertion, "Sadism demands a story, depends on making something happen, forcing a change in another person, a battle of will and strength, victory/defeat, all occurring in a linear time with a beginning and an end" (Mulvey 64, qtd in de Lauretis 1984, 103). De Lauretis draws the connection to narrative structure and takes this one step further, asserting that through narrativity itself, women are coerced into consenting to the patriarchal definition of femininity that upholds its oppressive power relations. The linear narrative that Aristotle placed at the center of western stages, and that has informed the quest structure of the online gaming MOOs, shapes complex and dangerous roles for women on the cyberstages of today.

In the trajectory of development that led to educational MOOs, Jim Aspnes's move to encourage expansion of the world over the quest for power and Pavel Curtis's break with narrative structure represent important, although not necessarily intentional, moves away from the patriarchal foundations of early online environments. However, the development of the code and programming language for MOO retained the inherently theatrical elements of character, location, and action. Identifying the importance of the theatrical metaphor in any virtual reality, Brenda Laurel's *Computers as Theatre* (1993) drew upon Aristotle's *Poetics* as a foundation on which to explore the design of human-computer activity at the level of one person and one computer, the interface that gives form and structure to the experiences human beings have with computers. Both technically and socially, the educational MOO environment operates as a metatheatrical space with implications far broader than those Laurel addressed,[1] a space in which the player/subjects create and enact characters that may or may not relate directly to the physical body or lived experience of the performer. This opportunity to perform in a potentially anonymous sphere distinguishes cyberstages from the physical stages of the live theater or film, television, and radio, where the players literally embody the characters they create. While the nature of the relationship between the technobody and the person who animates it remains one of the most hotly debated and exciting arenas for investigation, this essay will turn its attention to the theatrical constructs of MOO as a means to examine the potential of this text-based environment within an educational agenda.

This essay is divided into three sections, looking first at the theatricality of the MOO environment, then at the MOO I created for the Association for Theatre in Higher Education, and finally examining four different productions that have taken place at ATHEMOO. Exploring MOO as a theatrical environment necessitates an examination of the elements of MOO that highlight its specific performative nature, illustrating how the creation of rooms, objects, and players is distinctly theatrical. For example, a set designer and the creator of a room in a MOO both work to establish a particular atmosphere for the actions that will occur there; a costume designer and the person inhabiting a player object in a MOO both work to establish a look for the character that provides information for those who 'look' at it; and a playwright and the person inhabiting a player object in a MOO both work to create an identity for the character that is consistent through the use of speech patterns and actions. The section detailing the advent and administration of ATHEMOO serves as a case study in the creation of a MOO space intended to foreground the theatricality of the MOO environment while simultaneously shaping an equal-opportunity atmosphere for the exchange of knowledge and experimentation with the new stages cyberspace provides for theater makers. And while ATHEMOO was originally intended as a way to provide greater access to the organization's yearly conference, it has grown into a fertile space for cybertheater. What distinguishes ATHEMOO from other important professional and educational MOOs is the focus on theatricality and the shared concerns of its players regarding performance in cyberspace.

The final section of the essay examines four production events held at ATHEMOO as a means to illustrate the potential for intentionally theatrical activity in an already heightened theatrical environment.

MOOs as Theatrical Environments

Shakespeare's descendants might write, "All the MOO's a stage since all the men and women are player objects." The microcosm of each MOO, its particular flavor, appeal, and culture, is created not only by those who interact there, but by those who design, decorate, program, and organize the room objects. The theatricality of these objects distinguishes the environment from the literal enactment of daily life and establishes them as the stage set upon which the players interact and the play, as such, occurs. Generally, each educational MOO has a theme that serves to delineate the types of interactions that take place between those logged in at any given moment. While the themes of the earlier gaming MUDs located the patriarchal narrative quest structure as the foundation of interaction there, complete with a description of which obstacles one needed to overcome, educational MOO themes clearly eschew domination as a central concern and establish instead an environment for collaboration and learning. Logging in, one becomes or "animates" a player object, generally referred to as a *character.* Each character selects a gender, creates at least one description and set of interests, and by virtue of its class (player, builder, programmer, wizard) possesses a set of potential actions it can invoke. While these are, to a certain extent, remnants of the code structure of gaming MUDs, their existence within educational MOOs illustrates the theatrical nature of the environment.

Entering a MOO is not unlike entering a theater building. One travels to the site, a space that exists within society for the purpose of a specific activity that is both part of and apart from the real world around us. In theater architecture there is typically some sort of lobby or arrival area through which one passes to enter the theater by possessing a ticket. In cyberspace, one Telnets to the MOO location, arriving at a welcome screen that functions as an outer lobby through which one passes by connecting either as a player or a guest, in those moos which allow nonmembers to participate. Having connected, players and guests enter the first location in which interaction may occur or be observed, a secondary location that functions as an inner lobby from which to move around the MOO, or in the case of more established players, a room they have created and set as their home within the MOO. In either case, the arrival in a location that allows interaction marks the formal entry to the arena of the cyber theater.

In some ways, oddly enough, the cyber theaters of today are more similar to real theaters in Shakespeare's day than they are to theaters of the present. While current conventions of theater architecture dictate an audience formation that forces the full attention of all present to the performance, at the time Shakespeare wrote *As You Like It,* theater buildings had three divisions of the auditorium in which attention to the performance varied: the pit, the public galleries, and the private boxes. In the pit, for example, interaction between those in attendance was at least as important as the performance being presented. With the rare exception of productions staged in and for cyberspace at any given MOO, participants do not gather to watch a performance, but there are locations in the architecture of each MOO that have very specific conventions of interaction. For example, there might be a tutorial such as the one designed by Michele Evard for JaysHouse MOO, where several players are in the same room but do not interact, or there might be a coffee house such as Connections MOO's Tuesday Cafe, where open conversation is encouraged.

Although a theater building, as a whole, has as its central focus the performance space, it is a location that is divided into rooms with entrances and exits, specific functions and decor that informs the inhabitants of its purpose. The lobby, the auditorium, the dressing rooms, the green room where performers sometimes gather to wait for their entrances, all these are rooms to which and through which people travel. MOOspace is also organized into rooms, and like the set on the stage of theater auditoriums, each one is decorated in a manner that gives the audience information about the play to be performed. Because MOO is still a text-based environment, players read descriptions upon entry and imagine the space in which they are interacting. Nevertheless, the descriptions are generally quite vivid, detailing both the visual elements of the space as well as its feeling. One of the rooms at Meridian, a MOO environment devoted to the creation of "a global community shared by people interested in virtual travel, cultural interaction, friendship, and learning," is Keaonani's Luau, which upon entry is established this way:

> A large open space on Oahu's leeward coast, surrounded by glorious palms and coconut trees heavy with fruit. On your left is a large stage decorated with fabulous tropical foliage (no plastic here) and as the trade winds gently ruffle your hair, you smell the kalua pig roasting away in an imu. Fortunately, the Kona winds have died down, and the air is cool and fresh. Directly ahead of you is the most gorgeous beach, with incredibly fine sand and a beautiful view of the sunset past the

> rolling waves of the Pacific Ocean. To your right is the food preparation area, the tables are empty now, but soon will be filled with lomi salmon, baked sweet potatoes, macaroni salad, and poke.[2]

In addition, room descriptions in MOOs based on the LambdaCore include a line that indicates objects you "see" in the room, and some whole rooms are based on a widely used All-in-One room developed by Chad Wilson that allows sittables to be both indicated and used. The objects one "sees" and can therefore manipulate in a MOO room function like stage properties in the theater. Performers use properties to advance the action in a play, such as a briefcase from which to take documents or, as in the gaming environments, swords with which to do battle with an opponent.

The theatrical nature in the MOO—the rooms and objects as sets and properties—is further heightened by one of the most prominent remnants of the gaming MUDs, role-playing. However, in educational MOOs, this aspect of theatricality is divorced from Aristotle's linear plot structure and a patriarchal economy of relative values that places women only slightly above slaves. As Sherry Turkle, Amy Bruckman, Allucquére Roseanne Stone, and many others have noted, the separation of the virtual body from the physical body allows for incredible fluidity of identity on the cyberstage. In most theater performances, a group of actors assumes fictional identities, characters in the play to be presented. The relationship between the lived experiences of the performers and the characters they represent need not be, in fact rarely is, direct on the theater stage. The actor who takes on the role of King Lear in Shakespeare's famous play is unlikely to have experienced a functional monarchy, let alone been king in one. Yet this actor impersonates Lear through his lines, voice, and gestures. Once a person is given a password to access a player object in a MOO space, the creation of a character on the cyberstage begins with naming it, setting its gender, and describing it. This character may be directly based on the player's own life, as is common in professional and educational MOOs; it may be shaped around a famous figure from literature such as Lear; or it may be a fully fictional creation of the player's own making.

Characters on the theater stage wear costumes that illustrate their economic, social, and emotional status at any given moment during the play. As characters in the text-based MOO environment, players have several mechanisms with which they create costume. They incorporate articles of clothing and physical attributes into their descriptions, which are experienced by other players through the 'look' command. If a player looks at SteveS in ATHEMOO, the computer displays this description:

```
Brown hair with beginnings of distinguished grey at his
  temples and in his beard. Note the ponytail dangling
  backward from his receding hairline. He is dressed
  entirely in black from head to foot, with only a gold
  and silver pin on his chest.3
```

In dialogue, players may simply tell others what they are wearing or set their mood to reflect apparel. For example, Mary might use the @mood command so that when she speaks, the other players see:

```
Mary [stethoscope around her neck] says, "Time for a
  coffee break."
```

A further opportunity in the MOO environment to establish costume is to include it in the text that is displayed when players teleport into or out of a room. In each arena, character description, dialogue with other players, use of the @mood setting, or teleport messages communicate information about the character through the theatrical element of costume.

Since the physical bodies of players in MOO environments are not available to other players, the creation of a character's identity initially rests on the elements of costume and is extended through both speech and action. As on the theater stage, what characters look like or say about themselves is only the beginning. Enacting a fictional identity, both in the theater and in cyberspace, entails fashioning a consistent style of spoken interaction, a voice that establishes the character. For players in purely social MOOs or role-playing MUDs, the richness, coherence, and complexity of character is determined by the extent to which they wish to exert energy in playing the role. On these cyberstages, players can shape characters who bear no relation to their own social, economic, gender, racial, or sexual identities, characters who speak either very formally or frequently use colloquial expressions, characters who express themselves in short, terse statements or employ almost verselike poetry, characters who are warm and open or those who reveal very little about their feelings. Educational and professional MOOs differ in their policies about using real names and basing characters on real-life identities, but even where these MOOs do require full disclosure of such information, the presentation of self online remains within a performative realm.

While the physical bodies of the characters are literally absent from their onscreen interactions, MOO environments enable players to enact the body moving through space by way of textual descriptions of actions, or what in a play text is referred to as stage directions. Players on the cyberstage represent the movement of their virtual bodies with three different functions in the MOO, the 'emote' verb that allows them to describe what they are doing, verbs that provide shorthand access to common actions, and by their arrival and departure messages on the 'teleport' verb. While an actor on a theater stage can reach out to shake the hand of a colleague in different ways to layer the action with, for example, enthusiasm or reluctance, the MOO player does this through textual description. And just as theater actors modify their greetings depending on the situation and person they are greeting, these differences are expressed on the cyberstage through similar nuance in use of the 'emote' function.

The MOO environment exists as a sphere of theatrical interaction that is at the same time fictional (it takes place in an imagined environment between players who may or may not be enacting their actual identities) and potentially real (in educational and professional MOOs, meetings with real-life consequences frequently occur, such as Dene Grigar's 1995 dissertation defense that occurred at Lingua MOO—see chapter 9). While the activity in educational MOOs occurs in theatrical environments, the move away from the linear narrative quest structure that Aristotle placed at the center of Western theatrical activity establishes these locations as stages on which women have the opportunity to refuse patriarchal definitions of femininity. The description and programming of each room shapes it as a stage on which the action occurs. Without physical bodies to mark the players, they must enact characters using many of the same tactics theater performers employ. In fact, their performance extends beyond that of most stage actors because MOO players create their dialogue in the context of each exchange instead of Western theater's tradition of delivering memorized lines of text. On the MOO cyberstage, the theatricality of speech is reinforced by the way it appears on the screen, which looks quite similar to the play text on the page. Multiple elements of theater performance define the MOO environment and shape the activities that take place therein.

ATHEMOO, a Case Study

ATHEMOO is an educational and professional MOO that I created in June 1995 for the Association for Theatre in Higher Education (ATHE). ATHE is the major organization for theater scholars and educators in the United States, with over twenty-two hundred members. As ATHE's Vice President for Conference in 1995, I was in charge of the association's yearly gathering, which had been scheduled to take place at the Fairmont Hotel in San

Francisco. This event generally includes over three hundred sessions in a four-day period and is attended by roughly half of the organization's members. Fully aware of the fact that a large percentage of our members earn less than twenty thousand dollars per year, during the two years I spent in preparation for the conference I began to explore ways to provide access to those who could not, for a variety of reasons, attend the conference. Initially, I considered placing computers in the exhibit hall from which those in attendance could post reports on different sessions in the form of an unmoderated discussion list.

In January 1995, while casually perusing a magazine, I read about an Internet environment that I had never known existed, something called MOO that allowed for synchronous exchanges of conversation. Fortunately, I was able to attend a brief workshop in February organized by Judy Kirkpatrick and visited LambdaMOO, MediaMOO, and Diversity University. It was immediately clear that the MOO environment, with its simple commands for expression and opportunity to converse in real time with people around the world, was the perfect mechanism for expanding access to the ATHE conference. The ATHE board of governors, none of whom had ever visited a MOO, agreed that the MOO was a good idea for the expansion of the conference into cyberspace, and the University of Hawaii was willing to support the project by providing disk space. Thus, within a few short months and with no experience in computer programming, I had begun the process of creating ATHEMOO.

Although the original idea was that the MOO would function only as a conference-related venture, my work in developing the MOO lead me to believe that it was an educational environment of amazing proportions that could function within ATHE in numerous arenas, of which the conference was only one. In addition, I was convinced that I could build a community with a technologically friendly environment in which women could both accomplish goals related to their work in the theater and also attain knowledge and familiarity with programming. While the vast majority of players in the MUD environment are men, the population of educational MOOs is more balanced, as is the division between male and female archwizards. Despite the fact that my colleagues in the theater were almost completely unaware of MOO, I believed that the theatrical nature of the environment (as described above) would ease the resistance to this new technology. ATHEMOO would not only serve to accommodate the various functions of the national organization, it would provide an opportunity for both my female and male colleagues to learn how to use the technology for meetings, seminars, teaching, and performance in an atmosphere specifically designed for those new to MOO.

As a then untrained stranger in a strange land, I was fortunate to be introduced to Jan Rune Holmevik and Cynthia Haynes (creators of Lingua MOO) whose assistance in the design, educational features, and programming of ATHEMOO have been invaluable.[4] Contrary to my fears about being a woman in the traditionally male-dominated environment of computer programming, I have never felt denigrated as a woman or a newcomer. In fact, what I discovered on the stages of educational MOOs around the world was that there are several women in the top ranks of educational MOOs, beginning with Amy Bruckman, the creator of the first such MOO, MediaMOO. As I work to build ATHEMOO, men and women around the globe have assisted me with great generosity and patience, and my questions have never been met with derision or snide remarks about my gender.

The only resistance from the ATHE board of governors came from those who had read Julian Dibbell's article "A Rape in Cyberspace; or, How an Evil Clown, a Haitian Trickster Spirit, Two Wizards, and a Cast of Dozens Turned a Database into a Society" (1993) and, as feminists, were concerned about a potential lack of accountability in MOO that might make it an unsafe environment. Ironically, while the gender-specific violence described in Dibbell's article is to be avoided at all cost in any environment, the board's concerns about a MOO for ATHE might have served to limit women's acquisition of skill and knowledge there. In response to the board's initial hesitations, I explained that not having a MOO limited access to a sphere with great opportunities and that instead I could take measures to make it safer than LambdaMOO, where the rape in question occurred. In response to the board's concerns about potentially violent actions taken without full accountability, ATHEMOO policy was set to disable the automatic creation of characters and to require the use of real names within the MOO. While players do not have to be members of ATHE, their full names and email addresses are available for all registered players.

Since ATHEMOO began as a virtual conference center for theater scholars, teachers and artists, the first public spaces were modeled after a hotel lobby. In addition to taking steps toward accountability, one of my first steps toward creating a gender-balanced MOO was to name the guest characters after important figures from world theater history representing, unlike so many theater history texts, as many women as possible. In the time since ATHEMOO's first room was named the Lobby, Ken Schweller has brought in his Theatre, Zot O'Connor has brought in his Improv Room, Lee-Ellen Marvin brought in a tutorial for new members, and through the generosity of other programmers around the world we have an auditorium,

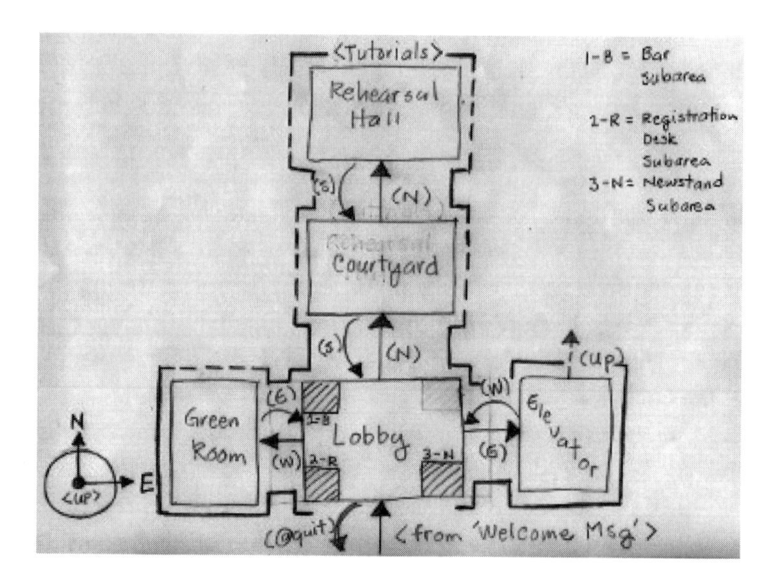

Fig. 11. Original drawing of the beginnings of ATHEMOO

a classroom and meeting rooms, an office suite, an inter-MOO communication system, and many new rooms, objects, verbs and features.

Performance Activity at ATHEMOO

Among the goals of ATHEMOO is to provide a site for experimentation with online performance. During the first year ATHEMOO hosted two performance events. With its inherent theatricality, the MOO environment is fertile for performance, and each project has utilized the technology in unique ways depending on the focus, MOO location, and performer-audience relationship of the event. In March 1996, Charles Deemer produced an online version of his hyperdrama, *The Bride of Edgefield,* in which simultaneous scenes occurred in four different locations. Later that month, Cat Hebert coordinated a brief performance in ATHEMOO's Ken Schweller Theatre in conjunction with the Crosswaves Festival in Philadelphia. Rick Sacks's *MetaMOOphosis* project, based on Kafka's novella, provides what he terms a MOO play-novel-improvisation environment. Since creating theater in the MOO environment does necessitate some programming skill, and since most MOO programmers are male, these first performances have all been undertaken by men. They range from performers delivering prescripted lines of text (for an audience that only reads along) to players who take on characters from Kafka's novel as a jumping-off point for improvisation.

Charles Deemer is an award-winning playwright who has spent the last ten years developing a form of environmental theater inspired by creating the scripts using hypertext. Five of these hyperdramas have been performed. Hyperdrama departs radically from linear narrative because instead of presenting a single story line with a beginning, middle, and end, different lines of the story happen simultaneously, and the audience must choose which to follow. Never performed in traditional theater spaces, hyperdramas are enacted throughout a performance space that allows the audience to experience different branches of the play in distinct, often distant locations. And, in order to view the entire production, spectators must attend the performance over the course of multiple evenings. Deemer's most recent production, *The Bride of Edgefield,* was performed twelve times in April, May, and June 1996 at the McMenamins Edgefield Bed and Breakfast Resort just outside Portland, Oregon. His newest effort, *The Last Song of Violeta Parra,* opened in March 1997 at the Prisma Art Gallery in Santiago, Chile.[5]

During the preparation for *The Bride of Edgefield,* Deemer discovered ATHEMOO and began to hold office hours there for his students. It was not long before the parallels between hyperdrama and the organization of the MOO environment led him to propose using the MOO for a test run of the production. The opportunity to create multiple rooms in ATHEMOO allowed him to evaluate the text in performance and to experiment with performance in the environment most closely related to the hypertext in which he had written the script. Player objects named after the characters appearing in the first fifteen minutes of the hyperdrama were created, along with four rooms dedicated to the performance that he named LabA, LabB, LabC, and LabD. The decision to limit the performance to the first fifteen minutes of the text allowed spectators who opted to stay for the duration of each scene to see all four parts in just over an hour.

Each player object's description was brief, giving information more about who the person was than a full description of the costume, body type, or mood of the character, and the textual description of the spaces in the labs were only phrases identifying the specific location of the scene. Whereas actors on the theater stage simultaneously use their voices to deliver lines and their body language to express emotions, during the performance at ATHEMOO, actors typed in their dialogue and were only infrequently able to use the 'emote' verb to intersperse the text with indications of physical action or feelings. "The hardest part," Deemer said in a recent interview, "was getting the actors technically proficient enough to *relax* and be able to improvise. This kind of theater always has moments of improvisation when timing of the script goes off."[6] Since moving between different locations in a MOO environment requires only one command, spectators at this early performance were able to (and did) move between scenes with greater frequency than previous hyperdrama performances. Unfortunately, due to a server crash, the performance was interrupted just before the fourth run of each scene was completed, and the discussion planned for afterward did not happen. However, undaunted by uncooperative technology, Deemer has begun thinking about writing a hyperdrama specifically for the MOO environment that will have a plot about people in a MOO.

In celebration of the fiftieth anniversary of the computer ENIAC, the American Music Theatre Festival launched Crosswaves, a five-year initiative linking new technologies and the performing arts. In March 1996, the Annenberg Center at the University of Pennsylvania hosted a wide range of events as the kickoff for Crosswaves. Cat Hebert, whose Virtual Drama Society works out of the Cyberloft Café in Philadelphia, organized a thirty-minute program to illustrate online performance that featured presentations in three cyber arenas, an IRC chat room, a two-dimensional graphical chat environment called the Palace, and ATHEMOO. For the performance at ATHEMOO, Ken Schweller installed his theater, which features a stage area, wings, auditorium, and rowdies balcony. The stage has a complex program that displays scenery descriptions at scheduled intervals so that the sense of place is not lost as the dialogue scrolls by.

The purpose of this performance at ATHEMOO was to demonstrate the potential of MOO environments for improvisational theater and what an improvisation would look like if one were attending online. Ironically, due to the necessity of illustrating three different environments in thirty minutes, Hebert prepared a script that his online performers delivered by means of prepared macros. Spectators onsite in Philadelphia watched a video wall while voiceover performers read aloud the text delivered by online characters to add an extra dimension to the performance. As per traditional theatrical convention, spectators online were invited to take a seat in the Ken Schweller Theatre Auditorium or Rowdies Balcony. After this segment of the program, ATHEMOO players and guests who had been online spectators met for nearly two hours to discuss the event and the issues it raised for theater on the cyberstage.

Both the hyperdrama and Crosswaves performances at ATHEMOO focused on a single event at which the online audience passively watched the delivery of prescripted dialogue on their screens. Canadian composer, musician, and author Rick Sacks's ambitious ATHEMOO performance project, *The MetaMOOphosis,* is designed to initiate improvisation around the characters and situations of Kafka's famous novella. This project utilizes sophisticated programming within a series of rooms to maximize the theatrical potential of the MOO environment. After a grand opening demonstration, *The MetaMOOphosis* space will remain in ATHEMOO for players to visit at their leisure. Because this project will continue to develop, the exact details of the space, costumes, objects, and programming may be different at the time this essay is read.

The performance space is defined by a series of eleven rooms, and a sign in the lobby notifies players to use the command '@go Kafka' to visit the site. Each room represents an important location of action in Kafka's novella, and they are joined not only by traditional MOO entrances and exits, but by keyholes that can be looked or spoken through. The first room is the Samsa Front Yard:

> `You stand outside the home of Gregor Samsa. He woke up a few mornings ago to find that he had been transformed during the night into a giant insect. There is a hospital behind the house and it looks like rain. Type 'enter' or 'in' to visit the house. Perhaps you can find a script and costume that will transform you as well. There is a closet in the foyer that may contain these items.`[7]

In the house there are three rooms on the first floor: the foyer, the kitchen, and the living room. Upstairs one finds a hallway with doors to the bedrooms of Gregor, Grete, the parents, and the three lodgers and of course, the stairs to Gregor's famous attic. The contents of each room includes objects inspired by the novella for the players to manipulate. For example, there is an apple core in the kitchen that can be thrown. If the apple core is taken into Gregor's room and thrown at his character, it becomes lodged between his shoulders and alters the tone of the lines in his script. Since the rooms in the performance space were built on Chad Wilson's All-in-One room, they are also programmed with sounds and smells that appear randomly and work to establish the mood of the environment.

In order to participate in an improvisation, players must select and wear a costume from the closet in the foyer. Players may also opt to simply move around the performance space and listen to those who have chosen to participate actively. Costumes for the characters Gregor, Grete, Mr. Samsa, Mrs. Samsa, Herr Doctor, and an observer are available in the closet, and each one has a script in the pocket. By wearing a costume, players gain the ability to use the 'say' command and speak either lines from the script or improvise lines of their own, while those who do not wear a costume are limited to using the 'page' command to communicate. Each script holds thirty to forty different lines either quoting directly from the novella, or written in the style of a Kafka work, complete with paranoia, sexual innuendo, and Freudian overtones, and accessed randomly (see Sacks). When players do not choose to improvise lines or use the 'emote' verb to establish physical action among themselves, activating lines of this scripted dialogue will create a performance.

Aliens and the Internet is the title of a series of four plays written by Steve Schrum, who teaches theater at Pennsylvania State University at Hazelton and manages the COLLAB-L discussion list. His ATHEMOO *Net-Seduction* performance project grew out of the final play in the series when he realized that performing it in a MOO environment mimicked an IRC chat room. Foregrounding the performative nature of identity on the cyberstage, part of the play's message is that "in the cyberworld you can be whatever you successfully pretend to be."[8] While this production is based on a script that the performers use for their dialogue, the script only provides a structure to keep the event moving forward toward a conclusion. Designed to incorporate improvisation, each performance holds the potential to be substantially different from the others despite the shared text.

The entire performance takes place in an ATHEMOO room, NetSeduction. Designed to operate like an IRC chat room, the opening screen contains both the room description and a Chat Daemon message. However, Schrum has designed a series of five adjoining rooms for private conversations and activities that might grow out of the performance event. Like the rooms in *MetaMOOphosis,* NetSeduction was built on the Wilson All-in-One room, and events like changes in music appear at intervals to set the mood of a disco bar. The mood is also enhanced by several robot characters placed in the performance space to deliver cliché lines of text and a menu board object on which players can write messages. The presence of the robots is both decorative and thematic, a comment on some of the styles of communication people in chat rooms employ.

Players may participate in three different capacities: players, supers, and lurkers. Players will take the six roles and be responsible for delivering the prescripted lines of dialogue. Supers will enact characters who hang out in the bar, chat with the audience throughout, and direct the action of the performance back toward the script after interludes of improvisation. One of the predetermined super characters is a bouncer, who both controls unruly spectators and reminds those who express discomfort with the adult language and situations of the production that they may leave at any time. Spectators are cast in the category of lurkers, those who watch but do not participate. In the event that there is more interest in the roles than there are scripted roles to play, Schrum will hold auditions to assign scripted roles to those possessing greater familiarity with the MOO environment and improvisational ability. Aside from the lag created by too many players in any given room, the number of supers is determined completely by the amount of interest expressed in the casting call. A series of rehearsals are planned, followed by three performances scheduled at times that accommodate audiences in Europe and Australia.

These four projects represent the broad range of production styles available to online theater artists experimenting with cyberperformance in the MOO environment, illustrating differing relationships to the theatrical traditions arising from Aristotle's patriarchal power structure. Hebert's presentation at the Crosswaves Festival kickoff most closely imitates traditional presentations of material with a beginning, middle, and end that audiences view as a single body. While Schrum's project is based on prescripted text presented in one location online, it is designed to incorporate exchanges of dialogue with the online audience as part of the presentation. Deemer's hyperdrama departs from traditional plot structure in its presentation of synchronous scenes in multiple environments, allowing the spectator to choose what to view and in what order. Like Schrum's *NetSeduction,* its premise includes potential interaction with the audience. Sacks's creation of an environment in which participants may take on a character to enact or simply watch as the action unfolds illustrates the greatest departure from traditional Western theater presentation while still maintaining links to aspects of it such as character, set, and dialogue. As illustrated in these theatrical productions on the cyberstage, there are MOO-dependent elements that represent new possibilities for the feminist subject: constructing gender identifications, a multiplicity of subject positions, and an opportunity for agency on the part of the audience.

Conclusion

At the center of its political agenda, feminist theory seeks self-determination for women, heretofore made impossible by a patriarchal heterosexual contract determined *for* women, binding us within terms and conditions assumed in contemporary society and forcing women into a "subject" position not of our own making. Rejecting the biological determinism of gender-based politics, a feminist subject, as a player on any of the world's stages, represents a "subject constituted in gender, to be sure, though not by sexual difference alone, but rather across languages and cultural representations; a subject en-gendered in the experiencing of race and class, as well as sexual, relations; a subject, therefore, not unified but rather multiple, and not so much divided as contradicted" (de Lauretis, *Technologies*, 2). Educational MOOs create a technical and performative arena with clear potential for the feminist subject.

Poststructuralist theory identified performance and technology as two of the most powerful forces that have structured society and our sense of the world in the twentieth century (Hawkes 56). The MOO environment is one arena in which the two intersect, inflecting one another so deeply that attempts to disentangle them prove nearly impossible. MOO is theatrical due to its basic technology through the remnants that remained at the level of code long after the narrative quest structures of the early gaming MUDs had been abandoned in favor of social or educational purposes. Performance on the theater stages where men and women are players is facilitated by the technology that operates tools used to build sets, illuminate scenery, or amplify sound. On cyberstages, the dependence on technology increases to the point where performance, in the form of speech and action, is only possible with the assistance of microchips and programming languages unfamiliar to the average member of a MOO community. But in conversation in a MOO environment, in the moment, my experience as a theater artist and scholar convinces me that first and foremost, the play's the thing.

Notes

1. Laurel's book remains an important contribution in contemporary thinking about human-computer interactivity in a dramatic context. Her Ph.D. in Theatre from Ohio State University and her many years in interface design uniquely position Laurel to examine the space of interaction between a human being and a computer program. What distinguishes her work from that undertaken in this essay is the introduction of multiple agents or human actors in the synchronous environment that MOO establishes.
2. This room, created by author Juli Burk, exists on Meridian, a MOO located at sky.bellcore.com 7777.
3. This is the character description for SteveS, a player object owned by Steve Schrum at ATHEMOO.
4. Special thanks are also due to Claudia Barnett, Jorge Barrios, Mike Copley, Isabel Danforth, Gustavo Glusman, Justin Granger, Diane Maluso, Lee-Ellen Marvin, Rui Mendes, Kevin Metz, Tom Riley, and Ken Schweller for investing large amounts of time, technical support, programming skill, and good humor in this venture.
5. The complete text for this hyperdrama is available at http://www.teleport.com/cdeemer/chile/chile-m.html.
6. Personal interview with Charles Deemer, August 1, 1996, ATHEMOO.

7. Room description from the Samsa House, a room object owned by Rick Sacks.
8. Personal interview with Steve Schrum, July 22, 1996, ATHEMOO.

Works Cited

Aristotle. *The Poetics.* Trans. Leon Golden. Tallahassee: University Presses of Florida, 1981.

Bruckman, Amy. "Identity Workshops: Emergent Social and Psychological Phenomena in Text-Based Reality." Masters thesis, MIT Media Laboratory, 1992.

Burka, Lauren. "MUDline." Available at http://www.utopia.com/talent/lpb/muddex/mudline.html (1995).

de Lauretis, Teresa. *Alice Doesn't.* Bloomington: Indiana University Press, 1984.

———. *Technologies of Gender.* Bloomington: Indiana University Press, 1987.

Dibbell, Julian. "A Rape in Cyberspace; or, How an Evil Clown, a Haitian Trickster Spirit, Two Wizards, and a Cast of Dozens Turned a Database Into a Society." *Village Voice,* December 21, 1993, 36–42.

Hawkes, Terence. *Structuralism and Semiotics.* Berkeley: University of California Press, 1977.

Laurel, Brenda. *Computers and Theatre.* Reading, Mass.: Addison-Wesley, 1993.

Mulvey, Laura. "Visual Pleasure and Narrative Cinema." In *Feminism and Film Theory,* ed. Constance Penley, 57–68. New York: Routledge, 1988.

Sacks, Rick. "The MetaMOOphosis Project." Available at http://www.io.org/rikscafe/kafka.html.

Schrum, Steve. *NetSeduction.* Available at http://www.hn.psu.edu/Faculty/Sschrum/RMTOCo/Netseduction.html.

Stone, Alluquere Rosanne. "Will the Real Body Please Stand Up? Boundary Stories about Virtual Cultures." In *Cyberspace: First Steps,* ed. Michael Benedickt, 81–118. Cambridge, Mass.: MIT Press, 1991.

Turkle, Sherry. *Life on the Screen: Identity in the Age of the Internet.* New York: Simon and Shuster, 1995.

IV

MOO Meditations

The New Circus prides itself on Being Real,
being nothing like a circus itself,
being something more like a house which nurses
images inside,

being something more like a house which welcomes
strangers, something more like a student watching
universes unfold from home. More like an
aurora rising

than the big top billowing a little plot.
Less particle than wave in particle heaven.
Further away than near, nearer than far away,
the tightrope walker

swims in language, waiting for direction. Give him
your address and The New Circus will come to you.
Between your voice's body and his body's voice
language buzzes.

—Brian Clements

11

Bodies in Place

Real Politics, Real Pedagogy, and Virtual Space

Beth Kolko

Traveling is always a matter of cross-dressing. The extent of the charade varies, but journeying from home necessarily entails either conscious or unconscious decisions about self-presentation. Do I want to blend in? Will I make myself less of a target for exploitation if I look like a *native?* Do I try to dress like the photos in the tourism books? For what local slang should I prepare? Or do I embrace my role as outsider and wear the visitor status proudly? Do I abide by the unspoken code that women in Brazil do not wear shorts in town? Or that women in Tokyo are only infrequently found wearing trousers? Do I wear the brighter colors of Paris? Or do I keep my clothes darker in color, more formal, as is common in Lisbon?[1]

When we journey from home, we pack a bag that, we hope, contains what is necessary for us to masquerade, as outsider or insider, in this new place. Whether it is an attempt to sand down the sharp edges that mark us as different, or a determined and more exaggerated display of who we are "back home," the decision of how to self-present while on the road is a fabulous puzzle. Whatever we decide, however, we are eventually forced to consider the relationship between body and place. What does my sense of self have to do with where I am? And how does changing where I am alter how I am interpreted by others? This precise question, the dynamic between the embodied self and geography, is thrown into question by the experience of traveling to virtual spaces, terrain unmapped and, in many cases, lacking detailed description of social structures.

Traveling to an online community is not altogether different from traveling to a new place about which one has failed to gather any substantive information. We can read accounts of chat spaces in popular magazines, we can heed the warnings of television reports, we can even eavesdrop on conversations among strangers who have ventured to these strange and disembodied places. But when we try to make the trip on our own, log into cyberspace and explore the barely marked doors, there will always be gaps between the local culture and our own. The gap between the known and the unknown is grounded, in such instances, at the site of the body. When traveling in real space, differences are played out on the terrain of the physical self. It is the clothes or the hair or the skin or the facial features or the height that marks one as definitively and immediately not-like. It is the body suddenly situated in unknown space that marks the experience; it is the contact of the self with the other that defines the tension of the moment. During a virtual journey, the dynamics do not change, but the contours of the gaps shift; edges blur, and categories fracture. In the course of a trip to cyberspace, the body is transposed, as if to another key, perhaps B minor, particularly jarring and unfamiliar to an ear accustomed to the gentle melodies of the supermarket, the classroom, the evening television programs. The sojourn in virtual space plays havoc with bodily senses, but, ultimately, it is this very journey that can provide us with a heightened sense of our selves in physical space, of our selves as agents, of our selves as embodied, partial, and political.

Writing the Politicized Body

This essay argues that learning to read the gap between the articulated, traveling, virtual self and the placed, familiar, physical self holds the key to envisioning the political possibilities of online worlds. How we physically present ourselves online is, in effect, a series of political choices. Similarly, how we choose to interact with the medium and other players is informed by another set of political decisions. The pages that follow attempt to describe the cadences of virtual interaction, both of the virtual self and the virtual community; ultimately, the self-presentation, contact, and conflict that define MOOs can be seen as political lessons, as each provides a way to think about identity politics.

Whether the online experience is a social or an educational one, characters interacting on a MOO eventually discover that they can play with their virtual bodies, that things are a little bit different in this place, that they can bend social codes and expectations and strictures. Not all characters act on this freedom, and not all use the potential for politically progressive ends. But the fact remains that sooner or later participants in a text-based virtual world come to recognize that their sense of self, of identity, is slippery.[2] The question arises—what then can we make of this realization? Is there a way to turn such play into pedagogically productive work? The argument here concerns setting students loose in MOOspace, whether an educational or social MOO, and waiting to see what happens as they realize how language and identity work in the medium.[3]

The textualized nature of the online self is the key to understanding why writing in a MOO is a form of political self-presentation. Character creation, and subsequent navigation and interaction, in a MOO can perhaps be seen as a form of autoethnography, wherein the writer explores both real and imaginary relations of power and culture.[4] The working through of versions of the self is a political project, and numerous theorists have explored this dynamic with relation to virtual space.[5] But the particular question of how identity is multiple in cyberspace is not the focus of this essay. Rather, I am interested in exploring how the situated body can intersect with identity and discourse in order to generate a viable politics from virtual interactions.

Situating the Body

The body occupies a central place in contemporary social theory; concerns with the body are, predictably enough, recurrent throughout work on cyberspace. Cyberspace theory overlaps with social theory, but it also departs in some substantial ways. When researchers write of cyberspace as a social space, of virtual communities as evocative of face-to-face communities, they draw attention to the codes of behavior that govern interaction in MOOs. Lynn Cherny and Lisa Nakamura, for example, have analyzed, respectively, the ways linguistic and racial codes circumscribe what can be said and performed online. Shannon McRae has written of how gender and sexuality are performed, and Sherry Turkle has discussed class and economic mobility as issues that inform characters' virtual interactions. The work of all these scholars talks around a central issue of online interaction: that is, all communication, all creation of the virtual space, occurs via text. The importance of language in this medium is an under-examined phenomenon. Much research has thus far focused on the vanished body in cyberspace rather than the embodied language of cyberspace. However, the intersection of these two issues provides a valuable area of inquiry. The body in cyberspace is a body out of physical space, but within discursive place. It is a body that is doubled, that is physicalized in "meat" space (in William Gibson's term) and narrated in virtual space. It is a placeless and yet placed body, embodied, situated, and yet partial and self-constructed. The body's partiality online is constructed, however consciously, by choices and negotiations made by the situated self. It is by increasing the embodied self's consciousness of the decisions made regarding the online self that the MOO becomes political space with transformative potential.

The doubling of the body in cyberspace and, in particular, its peculiar situatedness, is perhaps best examined in light of Donna Haraway's commentary on situated knowledges. In her essay of that title, as Haraway recapitulates the embodied history of women scientists and thinkers, she illustrates how the situatedness of bodies limits the ability of women theorists' formulations to be widely applicable (in the view of science).

She dwells on this tension for the duration of the essay, refusing to descend into an apolitical relativism, and instead recuperating the notion of the situated body by illustrating how a partial vision that is specifically grounded in the embodied self can provide a way to think through the questions of postmodernity without spiraling into an abyss of paralysis. She frames the central question in the following manner:

> So, I think my problem and "our" problem is how to have *simultaneously* an account of radical historical contingency for all knowledge claims and knowing subjects, a critical practice for recognizing our own "semiotic technologies" for making meanings, *and* a no-nonsense commitment to faithful accounts of the "real" world, one that can be partially shared and friendly to earth-wide projects of finite freedom, adequate material abundance, modest meaning in suffering, and limited happiness. (187)

As Haraway argues here, we need to be more than a repudiation of situated selves, even though that situated self is undoubtedly limited. While she wrestles with the nexus of relations surrounding issues of embodiment and vision, she is attempting to position the situated self within a site of political efficacy. The fracturing of so many categories under a condition of postmodernism liberates as a first gesture, but then constrains. The movement, enabled by a postmodern vision, from the margins to, if not the center, then at least a more central location, is an extremely powerful political move. However, this eddying of subject positions eventually evolves yet another internal current, and the patterns become ingrained, where constant movement never allows for the development of a political perspective. In the context of feminist theory both Haraway and Wendy Brown pinpoint the limitations of postmodernism for feminist efficacy. Brown writes of the condition of postmodernity:

> [P]ower reveals itself everywhere: in gender, class, race, ethnicity, and sexuality; in speech, writing, discourse, representation, and reason; in families, curricula, bodies, and the arts. This ubiquity of power's appearance through postmodernity's incessant secularizations and boundary erosions both spurs and frustrates feminist epistemological and political work: on one hand, it animates and legitimizes feminism's impulse to politicize all ideologically naturalized arrangements and practices; on the other, it threatens to dissipate us *and* our projects as it dissolves a relatively bounded realm of the political, and disintegrates the coherence of women as a collective subject. (70)

Brown's specific vision of fragmentation, which in turn obviates the possibility of movement, is central to the limitations of postmodern theory. And, in fact, it is postmodern feminists who have been at the forefront of recognizing the paralyzing aspects of postmodernism, and, in turn, working to recuperate postmodern theory for political purposes. The acknowledgment that partiality is limited as a political tool is what leads Haraway to argue so eloquently for another version of the subject-in-the-world. As she argues that we need to learn in our bodies, as she argues "for situated and embodied knowledges and against various forms of unlocatable, and so irresponsible, knowledge claims" (191), she posits this alternative to relativism precisely because it "is partial, locatable, critical knowledge sustaining the possibility of webs of connections called solidarity in politics and shared conversations in epistemology" (191). Her push here, then, is for a politics of situated partiality, an alternative to thorough embodiment or the free-floating commentator. Haraway's essay, written ten years ago, outlines a thorny and divisive question that continues to dominate much contemporary theory. However, her questioning is also prescient of emerging technologies and how these will affect theories of subjectivity. In her cyborg manifesto, Haraway addresses, in depth, questions of technology and biologism, but nowhere does her argument completely address the impact of synchronous communication technologies. What I would assert, then, is that her arguments of the last decade, via feminism, map out a series of questions that can help us think through the role of MOOs and MUDs in everyday life and in educational settings. Specifically, though, Haraway's formulation gives us a way to sees MOOs as social, educational and *political* spaces. That is, partial perspective as it relates to MOOspace would be the awareness that the virtual self is *constructed* on a variety of levels.[6]

The MOO as an Educational Space

Education, in this discussion, is taken as an implicitly political endeavor, and social space is also conceptualized as an always politicized space. How, then, can we discuss politics within a MOO, and how can we see MOOspace as a site of political possibility? Abstract theorizations of agency are too limited. It is imperative that we find ways to map the space of virtual communities as conceptual (but not virtual) spaces where participants are agents.

As the other chapters in this volume chronicle, the educational uses of MOO are varied. For both classroom and individualized instructional purposes, MOOs provide a space that encompasses play, conflict, conversation, and collaboration. Whether creating, describing, morphing, sending, or just speaking or emoting, interaction within a MOO forces users to manipulate a virtual self through amorphous space. The programming and narrative collaboration that eventually results in the built world of MOO revolves around members' positioning of themselves with respect to others. Each movement within virtual space necessitates conceptualizing the virtual body's relation to what surrounds. The rules of the physical world may be absent, but social codes remain. And whether participants choose to abide by or violate certain modes of behavior, or create new expectations and possibilities, the game that gets played is, ultimately, one of subject-construction.[7]

During MOOplay, participants self-constitute via words. As they construct an online identity, the slippage between the real and virtual selves varies meaningfully. On a MOO, participants mask gender, race, age, sexuality, nationality—any imaginable component of who we are can be bent (though not elided) in a MOO. Our bodies remain in the linguistic traces of what we type, program, and describe. But our bodies also remain situated at the site of the typist, the vision from that site partial to the character, and yet wider, able to imagine in the possibles of MOOspace. It is imperative that in this identity play, with the imagining of possible worlds and reactions and structures, we do not lose sight of the situatedness of the MOOing self. That is, part of the educational role of the MOO is to allow students to see how the grounding of their bodies plays out in virtual space; part of the real pedagogy is to examine the interaction of the self with geographical (dis)placement, and to see how language is used to manipulate a constructed self. Such a push heeds Brown's warnings, it is an embrace of Haraway's argument, it is an attempt to renegotiate with the claims that the cyberself is multiple.

In many ways, walking the thin line of virtual identity is the movement that Jennifer Mnookin makes in her arguments designed to discover a way to hold selves in cyberspace accountable. Mnookin, speaking from a legal perspective, wants something more than a fractured and multiple self set loose in virtual space. She wants bodies to matter, and she wants real consequences for virtual actions. It is, in fact, a feminist frame that would allow bodies to be both situated and able to speak to truth, to be both at loose in cyberspace and grounded in real space. That is, a situated stance does not preclude a relevance of vision, nor does the unsituatedness of MOO interaction obviate the importance of physical placement. Haraway reconciles these competing ideas with her new formulation of partial perspective and situated/local knowledges; we can use her schematic to interrogate the implications of MOOplay.

Haraway writes, "We need to learn in our bodies, endowed with primate colour and stereoscopic vision, how to attach the objective to our theoretical and political scanners in order to name where we are and are not, in dimensions of mental and physical space we hardly know how to name" (190). In other words, we have to understand what lies behind the self in cyberspace, that fractured and fragmented and multiple self chronicled by Allucquére Rosanne Stone, Sherry Turkle, Mark Poster, and so many others.[8] As Mnookin maintains, we need to be able to hold the self accountable, and current legal theory holds that only singular selves can be held responsible for actions. Similarly, Chris Schilling explores the relationship of the body to social theory and argues that we need a more complex theory of bodies than that allowed for by a strict social-constructionist perspective.

Shilling's synthesis of various social theories is, in effect, a mirror of Haraway's claim that there must be a way to view the self as situated and yet partial. He remarks, "To begin to achieve an adequate analysis of the body we need to regard it as a material, physical and biological phenomenon which is irreducible to immediate social processes or classifications" (10). Shilling's assertion with respect to how the body walks the line between embodiment and partiality is particularly germane to a discussion of the dynamics of MOO action and interaction. Shilling thinks "the body is most profitably conceptualized as an unfinished biological and social phenomenon which is transformed, within certain limits, as a result of its entry into, and participation in, society" (12). Again, this particular duality mirrors the movement of the self into virtual space. That is, as a typist manipulates a MOO character, the body that moves through cyberspace is a biological phenomenon sitting at a keyboard, and a political phenomenon that has gained access to the technology, and a social phenomenon that is interpolated by the social conventions of the typist's environment and the MOO character's virtual world. Taken together, that is a significant amount of construction. The choices made throughout virtual journeys are grounded in the physical world and influenced by the constructing forces there. The MOO's power lies in our ability as educators to make students aware of how the characteristics of their situated selves inform their virtual decisions. The *layers* of construction within this process are of particular relevance. That is, there is the construction at the level of the social world *and* the virtual world, processes of social constructionism interacting with the unfinished biological phenomenon that is the flesh. A consciousness of these layers of construction is a political consciousness.

When, for example, cyberspace commentators invoke Gibson's notion of the body as "meat," they are acknowledging that there is some separation between the seated, situated self and the virtual, formless self. The meat, however, is not the only example of a situated self in online interaction. I want to repeat this claim here, because it is a complex one, and also central to the overall argument of the chapter. The situated self of cyberspace is subject to varying levels of construction and constriction. The social environments of meat space affect the embodied typist, and the social conventions of cyberspace impinge upon the matterless form of the virtual self. Classroom practice that uses the MOO to illustrate how the different selves are provided specific choices at some levels, and denied other sorts of options, is embedded in a political pedagogy.

The MOO as a Narrative Environment

There is a particular cadence to MOO interaction that evokes nothing so much as a narrative collaboration. As Lynn Cherny, Amy Bruckman, and others have argued, participation in a MOO is a series of interactions with other characters and with the programmed/programmable database that constitutes the virtual world. Thus, even when I am alone in a room, looking at objects, or reading MOO-internal mailing lists, I am collaborating, albeit asynchronously, with other players/programmers. We can consider this collaboration much in the same way we consider the narrative collaboration of two or more MOOers who narrate a story to one another, building upon the others' contributions. The dialogic components of such communication provide another way of conceptualizing the political potential of MOOplay. That is, through the combination of viewing from one's perspective, and interacting with another perspective, a politically viable partial perspective merges. Thus, it is by using MOO to examine the reified relations of identity and power in real space, precisely by viewing the slippage of such relations in virtual space, that the educational use of MOO as an object lesson in politics becomes most articulated.

Current work in narrative theory provides one way of examining the stakes of collaboration, and hence the stakes of narrative choices characters make in online collaborations. In particular, scholarship that emphasizes the ethical obligations inherent in narrative exchanges, although not directly related to the argument about political potential, does provide a sense of why collaborative choices might matter in the political sphere. For example, when Adam Newton writes about narrative ethics, one of his central arguments focuses on how the teller and the listener are connected in an ethical relationship. What I would draw from Newton's argument is the implied power relationship of a narrative collaboration. MOO is about communicative acts that are collaborative. By extension, then, the MOO is going to function as a playing field for the kinds of ethical relationships Newton lays forth. While ethics are not the center of my argument, the ways in which Newton evokes power and accountability *does* have relevance here. The ethics of narrative collaboration are concurrent with the exploration of identity and place, for in a MOO, all such exploration occurs via narration, and all negotiate power relationships. Similarly, the self-presentation in a MOO is an exercise in partial and situated knowledge. The virtuality of MOOplay does not obviate the political implications inherent in the work done online. The MOO is, finally, a place of contact and conflict, both with the self and with others. While Haraway's situated knowledges map the best way to understand that constructing a partial and virtual self is an exercise in politicized self-articulation, narrative theory provides a strong background for understanding the contingent nature of synchronous interaction online.

Narrative theory highlights the significance of narrative collaboration online; current work on contact zones provides a similar gloss to the intersection of self-presentation and power. Mary Louise Pratt presents contact zones as a way to enable politically viable exchanges. Pratt's contact zone is a site of conflict, but also of connection. The conflict in her formulation arises from the gap between communicative styles and the situatedness of participants. Pratt uses contact zone "to refer to social spaces where cultures meet, clash, and grapple with each other, often in contexts of highly asymmetrical relations of power" (34). She extends this term to include classroom communities, and I would extend the following argument to include explicitly educational MOOs. The contact zone Pratt outlines is one in which the friction of contact provides the heat for political action. She focuses on a particular class at her institution, one that was refashioned from classics-of Western-civilization to what Pratt calls "a much more broadly defined course called Culture, Ideas, Values." She describes the class in the following way:

> In the context of the change, a new course was designed that centered on the Americas and the multiple cultural histories (including European ones) that have intersected here. As you can imagine, the course attracted a very diverse student body. The classroom functioned not like a homogenous community or a horizontal alliance but like a contact zone. Every single text we read stood in specific historical relationships to the students in the class, but the range and variety of historical relationships in play were enormous. Everybody had a stake in nearly everything we read, but the range and kind of stakes varied widely. (39)

Pratt's final assessment of the course is that, by its very nature, it "put ideas and identities on the line" (39). This particular effect is what I would focus on in respect to the MOO.

The MOO as a Contact Zone

Inasmuch as there are power, status, and cultural differences among the population that manages to get itself online, the MOO is a place "of unsolicited oppositional discourse, parody, resistance" (Pratt 39); the MOO is a place where the collisions of worldviews, of language, of cultural expectations are always in motion. Both the process of self-presentation and the clashing of individual users on a MOO creates a version of Pratt's contact zone. Earlier in this chapter, the relationship between bodies and virtual space was discussed, and Haraway's conception was put forth as a way to map the kinds of identity play that occur in cyberspace. Pratt, in her dealings with the clash of identities in the contact zone, extends the conversation. Her formulation lays the groundwork for conceptualizing the MOO as a contact zone within which players are constantly negotiating aspects of their identity, gender, race, and age. In so doing, they are creating a space within the MOO that can be as political as Pratt's refashioned classroom space. In this sense, the MOO can be seen as effective as both an official and unofficial educational space.

Pratt's contact zone is a site of narrative collaboration. The conversation is a cultural one, but the construction of stories is at the center. What causes the tension is the struggle over who will dominate the narrative, or at least be able to have some effect on its outcome. This process is dialogic, and real political consequences are at stake. I would argue that part of Pratt's goal in teaching the class is to make the conversation more dialogic. That is, she seems interested in increasing the number of contributors to the cultural (hi)story. Dialogic processes hold political potential for Pratt. They hold similar promise for Haraway. They are, finally, one of the reasons I see so much possibility in MOOplay.

The processes of self-representation on a MOO must contend with a variety of layers of social constructionism. The identity (or identities) a student adopts on a MOO is chosen for complex reasons that intersect with ideas of the possible, the probable, and the appropriate. The narrative collaborations of that same student hinge on a similar series of negotiations. These maneuvers through language, the attempts at linguistic representation, are a game, reminiscent of the language games of Jean-Francçis Lyotard. They are games that can be contextualized as part of a practical political agenda, particularly the kind Lyotard aims for as he wrestles with the role of language in the search for justice. Such a consideration of justice is not altogether theoretical, not for the MOO, and not for Lyotard. When he claims, for example, that the ability to judge is not dependent on criteria, that the form "is that of the imagination. An imagination that is constitutive" (17), he invokes a venue and process strikingly similar to the MOO. He goes on to describe this imagination as "not only an ability to judge; it is a power to invent criteria" (17). In so doing, Lyotard seems to be describing the MOO, a place where imagination and language are constitutive, creative, a place where learning to judge how power constrains choices can generate lessons valuable for the physical world.

The determinations that inform the narration of the virtual self, and the contents of the virtual world, are politically coded. Decisions about race, class, geography, and gender are informed by cultural codes. Movements in narration, room description, and online storytelling reflect similar cultural values and power relationships. Narrating the self and the contents of virtual space reflects the situated self's struggle with the political world. Slippage and movement and negotiation provide a world of possibles, a world our students are perched on the verge of exploring. The conversation and narrative collaboration of MOO interaction does not simply invoke the ethical bond between narrator and narratee; it calls forth an embodied and yet socially constructed self; it is the constant enactment of the contact zone with the attendant discoveries about the self and other; it is predicated wholly on discursive exchange. MOOing is a matter of negotiation through language, of navigating the virtual body through virtual space, and yet positioning that virtual self, existing as embodied meat and disembodied fluidity. What looks so much like play online, then, is much more than diversion.

Haraway concludes "Situated Knowledges" with references to the coyote, the trickster, to humor's ability to tweak us, to rob us of the myth of mastery. The MOO, after all, is the domain of the coyote; the MOO is a realm of play, of pun, of slippage. Her final assertion that it is "the world as coding trickster with whom we must learn to converse" (201) is the argument that the MOO is the space within which we must learn to move, think and talk. The particular combination of work and play, of the virtual and the real, of stakes measurable and amorphous gives the MOO its potency as an educational space. Ultimately, the MOO with its lessons in play, partiality, and situatedness holds a particular power to teach us about our bodies, our place, and our politics.

Notes

1. These examples here are drawn from social codes governing women's dress, a convenient place to look for variously articulated and enforced rules.

2. It's not my intent to totalize MOO interaction with this portrait of online behavior. Clearly, MOOplay does not lead all participants to see their own identity as slippery. And although some who do not see themselves as implicated will observe other participants' reactions to being online, there are those who will completely resist the "fractured identity" interpretation. I suspect it is impossible to account for every possible reaction to the MOO; what I hope to provide here is a generalized tale of life online.
3. When I discuss MOOs here it is not within a context of using MOOs to approximate the teaching strategies of distance education. Similarly, I am not addressing MOOs used as a replacement for LAN-based collaborative software.
4. The notion that writing in a MOO is autoethnography is an idea for more detailed exploration elsewhere. However, it seems clear that the process of presentation involved in creating online characters and geography entails a discursive working out of the self that is similar to the self and cultural critique of autoethnographic writings.
5. See the work of Sherry Turkle, Allucquére Roseanne Stone, Mark Poster, Heather Bromberg, Lisa Nakamura, and Shannon McRae for further explanation of this dynamic.
6. Another way of phrasing this is that selves in MOOspace and in real space are virtual. The importance of MOOplay is the way it forces participants to confront the discursive constructedness of their identity (scripts, if you will).
7. The social codes that different participants bring to and play out on MOOs vary widely. Similarly, it would be difficult to identify one set of codes that are enforced or valued uniformly on any particular MOO (like the face-to-face world), and the divergence across MOOs is even sharper. The gaps in the codes brought in by players, and the clash of competing unwrittens, provide precisely the political potential this essay addresses.
8. These critics have been at the forefront of those detailing the particular way cyberspace fragments subjectivity. The argument, in summary, is that the ease with which an online character can change the representation of the physical self leads to self-exploration and, ultimately, discovery of how components of one's identity effect larger interactions. In turn, then, users of online media are able to see the constructed nature of what they think of as their "self," and they discover a freedom to experiment and explore otherwise suppressed elements of their selves.

Works Cited

Bailey, Cameron. "Virtual Skin: Articulating Race in Cyberspace." In *Immersed in Technology: Art and Virtual Environments*, ed. Mary Anne Moser with Douglas MacLeod, 29–50. Cambridge: MIT Press, 1996.

Bromberg, Heather. "Are MUDs Communities? Identity, Belonging and Consciousness in Virtual Worlds." In *Cultures of Internet: Virtual Spaces, Real Histories, Living Bodies*, ed. Mark Shields, 143–52. London: Sage, 1996.

Brown, Wendy. "Feminist Hesitations, Postmodern Exposures." *Differences: A Journal of Feminist Cultural Studies* 3, no. 1 (1991): 63–84.

Bruckman, Amy. "Programming for Fun: MUDs as a Context for Collaborative Learning." 1994. Available at ftp://media.mit.edu:/pub/asb/papers/necc94.txt.

Cherny, Lynn. "Objectifying the Body in the Discourse of an Object-Oriented MUD." *Works and Days* 13, nos. 1–2 (1995): 151–72.

Haraway, Donna. "Situated Knowledges: The Science Question in Feminism and the Priviledge of Partial Perspective." In *Simians, Cyborgs, and Women: The Reinvention of Nature*. New York: Routledge, 1991.

Lang, Candace. "Body Language: The Resurrection of the Corpus in Text-Based VR." *Works and Days* 13, nos. 1–2 (1995): 245–60.

Lyotard, Jean-Franqis, and Jean-Loup Thebaud. *Just Gaming*. Trans. Wlad Godzich. Minneapolis: University of Minnesota Press, 1985.

McRae, Shannon. "Coming Apart at the Seams: Sex, Text, and the Virtual Body." In *Wired Women: Gender and New Realities in Cyberspace*, ed. Lynn Cherny and Elizabeth Reba Wiese, 242–64. Seattle: Seal Press, 1996.

Mnookin, Jennifer. "Bodies, Rest, and Motion: Law and Identity in Virtual Spaces." Cyberlaw Session. Virtue and Virtuality: Gender, Law and Cyberspace Symposium. Massachusetts Institute of Technology, April 20, 1996.

Nakamura, Lisa. "Race in/for Cyberspace: Identity Tourism and Racial Passing on the Internet: The Resurrection of the Corpus in Text-Based VR." *Works and Days* 13, nos. 1–2 (1995): 245–60.

Newton, Adam Zachary. *Narrative Ethics*. Cambridge, Mass.: Harvard University Press, 1995.

Poster, Mark. "Postmodern Virtualities." In *Cyberspace/Cyberbodies/Cyberpunk: Cultures of Technological Embodiment*, ed. Mike Feathersone and Roger Burrows, 79–96. London: Sage Publications, 1996.

———. *The Second Media Age*. Oxford: Polity-Blackwell, 1995.

Pratt, Mary Louise. "Arts of the Contact Zone." *Profession* 91 (1991):33–40.

Shilling, Chris. *The Body and Social Theory*. Thousand Oaks, Calif.: Sage Publications, 1993.

Stone, Allucquére Roseanne. *The War of Desire and Technology at the Close of the Mechanical Age*. Cambridge, Mass.: MIT Press, 1995.

Todd, Loretta. "Aboriginal Narratives in Cyberspace." In *Immersed in Technology: Art and Virtual Environments*, ed. Mary Anne Moser with Douglas MacLeod, 179–94. Cambridge, Mass.: MIT Press, 1996.

Turkle, Sherry. *Life on the Screen: Identity in the Age of the Internet*. New York: Simon and Schuster, 1995.

(Non)Fiction('s) Addiction(s):
A NarcoAnalysis of Virtual Worlds

D. Diane Davis

> ***Two unequal columns, they say distyle [disent-ils], each of which—envelop(e)(s) or sheath(es), incalculably reverses, turns inside out, replaces, remarks, overlaps [recoup] the other.***
> —Derrida, *Glas*

> ***To resist electronic technology is as futile as trying to turn back the tides. It has already swept over us in ways we have yet to realize. It is not a question of whether to accept or reject this new world but of who is going to use it and how.***
> —Taylor and Saarinen, *Imagologies*

Pre-(R)amblings:

The assumption that there exists a rigid distinction between virtual reality (VR) and "real life" (RL) prompts a series of concerns for educators, not the least of which is whether there can be any "real" value gained by doing time in a "virtual" world. Certainly, the number of universities officially banning the use of their mainframes for MUDding (as well as Australia's continent-wide prohibition on the activity) attests to this concern. Some cyber enthusiasts have attempted to answer it by noting that several text-based virtual realities now provide primarily educational rather than gaming environments. This book, for instance, is devoted to explaining the ways in which educational MOOs invite "players" to engage productively in online synchronous academic conferences, discussion groups, even dissertation defenses. That "playing" in a virtual world can result in real-life learning should be of little doubt by the end of *High Wired.* The question still being begged in this compelling answer, however, is whether there is indeed a clear distinction to be maintained between the "self" and the techno-body, the simulacrum and "The *Real* Thing," small talk and big thought.

As we move toward the end of the second millennium and into a post-almost-everything era, it becomes increasingly difficult to justify such distinctions. In fact, Jacques Derrida and Jean Baudrillard, among others, have called our attention to the ways in which "truth" and "representation" operate as functions of play and simulation, and that certainly mucks things up a bit. But it also opens up some interesting possibilities for *linkage* and *leakage*. We will set our scanner to those possibilities and suggest, among other things, that what educators hold against VR is roughly equivalent to what the state holds against illicit drugs: they both challenge the reality principle. The techno-body and the drug addict escape into a world of simulacra, and this is unacceptable, as Derrida notes in "The Rhetoric of Drugs," because "[w]e disapprove . . . of hallucinations" (7). We're uncomfortable because their virtual pleasure takes an undeniable swipe at *our* reality, a swipe that suddenly blurs the border zone between VR and RL, and therefore between me and my techno-self, my work and my play, my big thought and my small talk.

What it will have been necessary to admit is that RL is as much a function of VR as it is the other way around. The cycle of reproduction, according to Mark Taylor and Esa Saarinen,[2] goes like this: "knowledge produces technologies that produce knowledge" ("Interstanding" 6). Educational MOOs invite connections between RL and VR that leave our understanding of both profoundly altered. It makes little sense at this point to refer to VR as *fake* reality. There is no reason, as Sherry Turkle notes, to make VR and RL compete for the status of the *real* (236). Inasmuch as VR, as Avital Ronell suggests, is about "Being-in-the-world, and liberating the location of Being to non-substantial spaces, it is trying to reconfigure the possibilities for sharing the world" ("Activist" 299). VR, then, offers us not a *fake* experience but ***another inflection of Being***—an inflection that shows up somewhere between low life and high theory, rumor and philosophy. As such, it deserves validation as simply *an/other* reality, one that is intricately connected, full-duplex power, to *where we live*. But make no mistake, this validation will cost us something. A reality that fluidifies the border between production and seduction will shake—*is shaking*—the foundations of the academy.

We've known for a long time now, at least since Heidegger and de Man, that though ph[A]losophy puts on the ritz from the heights of the Ivory Towers, it also hangs out in the back alleys of some mighty shady neighborhoods.[1]

We cannot feign to begin with the chronological beginning, pretty much with The Life of Jesus: there's no sense in privileging here the law of temporal or narrative unfolding that precisely has no internal and conceptual sense.
(Derrida, *Glas* 5)

We won't pretend to build an argument here; we will proceed other/wise, on the "basis" of a different sensibility, toward something other than closure. Our primary interest here lies in three re/cognitions: (*a*) that there is no Being that is not Being-*on-something*; (*b*) that even the fantasy we call "reality" is a function of simulation; and (*c*) that MOOspace exists in the in-between of some of the West's most fundamental dichotomies: fantasy and reality, organism and machine, and seduction and production. We will offer some [dis]connections among these points, shooting not so much for a reasoned argument (though, something like splotches of argument will show up in this text) as for the sounding of an/Other knell and a tuning into that which has too frequently gone unheard.

Better buckle up. *We'll*
have
had
no
choice
but to ssspeed.

Being-*on-Fictions*:

Derrida asks, What do we hold against the drug addict? His answer: Something that we do not hold against the smoker or the alcoholic: that the druggie "cuts himself [*sic*] off from the world, in exile from reality, far from objective reality and the real life of the city and the community; that he escapes into a world of simulacrum and fiction. We disapprove of his taste for something like hallucinations" ("Rhetoric of Drugs" 7). The drug hysteria of the 1980s that spawned the pathetic "Just Say No" campaign taught us that "drugs make us lose any sense of true reality." It's not, Derrida says, that we "object to the drug user's pleasure per se, but we cannot abide the fact that his is a pleasure that is taken in an experience without truth" (7–8).

In chapter 10 of Plato's *Republic*, Socrates argues that true reality, closest to the God who created it, exists in the ideal forms, from which the "reality" we see models itself. Tables produced by carpenters are representations of some ideal table. The tables we have in our homes, then, according to Plato's

What do we hold against the drug addict? The same thing Plato held against poets and painters: that the pleasure they enjoy is "inauthentic" (8). Both the poet and the druggie are creatures of the simulacra; they enjoy a sham world twice removed from "true" reality, a world of seduction and illusion. For this, they are "driven from the community by Plato[nists]" (8). Interestingly, what we hold against net-heads in MOOspace is the very same thing: that they spend their time in a fictional world, in a "consensual hallucination," a world devoid of truth and "far from objective reality." What we good Platonists hold against the poet, the drug addict, and the net-head is that they prefer to shoot up fictions, to live in the world of appearances and seduction rather than truth and production.

There can be little doubt, as Taylor and Saarinen suggest, that "Virtual Reality is the LSD of the electronic age" ("Addiction" 3).[3] Anything can function as a drug, and certainly these days VR and other cyberprojections figure among our favorite pills to pop. The Net, they say, "is something of a narcotic. It's addictive—terribly addictive. Surely it's possible," they note, "to O.D. on the net" (4). VR, even text-based VR, has proven itself to be one potent trip. If you're not convinced of this, spend a little time reading the posts of the newsgroup called "alt.mudders.anonymous." That should do it. MOOing is tripping. And yet, if we're tempted to start spouting Nancy Reagan–speak, to condemn cyburbia as something to which we should (or could) "Just Say No," it's time for a rigorous hesitation.

.

.

We shouldn't be too hard on ourselves, of course, because *most* of us are

Socrates, are second-order representations, once removed from true reality. Poets and painters, on the other hand, represent these representations; theirs are, therefore, third-order representations, twice removed from true reality. According to this setup, MOOspace would be a third-order re/present/ation, as well, since it both represents physical representations (universities, old abbeys, dungeons, cities, etc.) and invites the techno-bodies that inhabit those representations to re-present any number of "actual selfhoods." No doubt about it, Plato would freak if he were around to see this.

But he's not; in fact, by now his supposedly enduring setup looks embarrassingly dated. Look around. The

nursing a humanist blind spot the size of a Mack truck. But it's time—is it not? as the end of the second millennium approaches?—to wean ourselves from such comfy naïveté. What wants to be said here, what demands a hearing in the face of this inclination toward prohibition is that, at some level, we've always already said yes:

There
are
no
drug-
free
zones

The whole show (of Be[com]ing) comes down to questions of dosage and mixture; there are, to be sure, good and bad trips, but there is no authentic Being that is not always already Being-*on-something*. Even the language we speak also (more so) speaks us, operates less as our possession than our possessor; or, if you prefer, our dis-possessor. As Ronell—who had been reading Heidegger, who had been reading Nietzsche—suggests, it is now difficult to deny that a certain kind of Being-on-drugs is fundamental to Being itself (*Crack Wars*).

In *Crack Wars: Literature, Addiction, Mania*, Ronell demystifies our "average-everyday" understanding of "drugs" and "addiction," redescribing both from an "excentric" angle, from the angle of the possessed. She begins by generalizing our notion of "drugs," suggesting that we have been duped by and about them, in part, because we have expected

assumptions upon which Plato based his philosophy of representation are no longer tenable. Both God and Man have been killed off. All foundational theories have proven themselves to be still under (a dead) God's thumb. Imagology has all but shattered ideology's dignity. There can be little doubt that we have moved on.

Jean Baudrillard offers a reconfiguration of Plato's setup for the electronic age; he presents another look, in the late twentieth century, at the notion of representation by working through an analysis of the "successive phases of the image." One might assume that the image is:

1. the reflection of a profound reality; or that
2. it masks and denatures a profound reality; or
3. it masks the absence of profound reality; or
4. it has no relation to any reality whatsoever; it is its own pure simulacrum. (Cf. Baudrillard 6.)

These phases mark a transition, Baudrillard says, from "signs that dissimulate something to signs that dissimulate that there is nothing," and this, he says, is a "decisive turning point." He continues:

> **The first reflects a theology of truth and secrecy (to which the notion of ideology still belongs). The second inaugurates the era of simulacra and of simulation, in which there is no longer a God to recognize its own, no longer a Last Judgment to separate the false from the true, the real from its artificial resurrection, as everything is already dead and resurrected in advance. (6)**

them to operate within a "restricted economy." We have assumed that one is either on drugs or not on drugs and that Being is good but that Being-on-drugs is bad. We have assumed an either/or structure and, therefore, the capacity to make an easy and clear distinction between the two. This logocentric formulation of "the drug problem," Ronell suggests, has led us into an ugly war on drugs, which is proving to be unwinnable. But Ronell offers an/other formulation of the problem: perhaps we have waged war with a structure (Being-on-drugs) that operates as a fundamental tenant of our existence? What if "drugs," Ronell asks, were not simply "one technological extension among others, one legal struggle, or one form of cultural aberration?" What if, instead, "'drugs' named a special mode of addiction . . . or the structure that is philosophically and metaphysically at the basis of our culture?"

Ronell notes that the dis/covery of endorphins marked the moment at which it became possible to recognize drugs as "excentric," as an outside that is also always already inside: "Endorphins relate internal secretion to the external chemical" (*Crack Wars* 29). Marking language as another excentric "drug," Ronell notes that when we borrow the words of others to make ourselves understood, we enter the realm of what could be called Being-on-their-language. When Benjamin quotes Baudelaire to explain what he feels when hashish begins to take effect in his own body, he "takes an injection of a foreign body (Baudelaire's *Les Paradis artificiels*) in order to express his inner experience." And Ronell reminds us that this "is by no means an atypical gesture. To locate 'his' ownmost subjectivity, Thomas De Quincey cited Wordsworth." These texts," Ronell says, "are on each other. *A textual communication based on* ***tropium***" (29, emphasis added).

Tr/opium, like opium, is a mind-altering, habit-forming drug—we can't get through the day without it. What Ronell calls "tropium" brings to mind Derrida's definition of writing in *Positions*: "extraction, graft, extension" (71). The condition of possibility for the coming into Being of any text/identity is its capacity to graft itself onto some other text/identity. And here, the easy binary between **Being authentically** and Being-*on-**drugs*** breaks down/up. We have to be *under the influence* (of language, of technology, of History/Tradition), Ronell suggests, to Be at all. If drugs are "mind-altering" or "frame-setting" substances, we are high or low on them all the time. Being-on-drugs indicates, she says, that "a structure is already in place, prior to the production of that materiality we call drugs, including virtual reality or cyberprojections" (*Crack Wars* 33).

"Drugs," Ronell says, "are crucially related to the question of freedom" (*Crack Wars* 59). But what is freedom when to Be at all is to be hooked, to be an addict? Liberation, it seems, must be re/thought. "Drugs do not," Ronell says, "close forces with an external enemy (the easy way out) but have a secret communications network with the internalized demon. Something is beaming out signals, calling drugs home" (51). To be a human-Being is to be dependent for one's very life on something that is *not human* but that **brings the human into Being**. There is no Being that is not possessed, thrown, animated by one "drug" or another. "Reality" itself is a "consensual hallucination," an effect, a function of other functions. What Plato was hiding is what poets and painters, as well as some net-heads and drug-heads, happily embrace: that the so-called third order of representation may be *as true as truth gets*. The *fantasy* we like to call "reality" may be always already, as Baudrillard insists, a function of simulation.

The Cycle of Simulation:

"[W]e are moving from a modernist culture of calculation toward a postmodern culture of simulation." (Turkle 20)

Baudrillard begins *Simulacra and Simulation* with a look at "the Borges fable in which the cartographers of the Empire draw up a map so detailed that it ends up covering the territory exactly." This reproduction is so realistic that eventually, as the map frays and the Empire declines, the simulation is mis/taken for the real thing. With a kind of Promethean nod, Borges relates a fable in which a simulacrum is confused with Truth/Reality/the Real. But today, Baudrillard notes, simulation is no longer a matter of copying an original. Rather, today, simulation is "the generation by models of a real *without origin or reality:* a hyperreal" (1). Baudrillard continues:

> **The territory no longer precedes the map, nor does it survive it. It is nevertheless the map that precedes the territory—precession of**

> **simulacra—that engenders the territory, and if one must return to the fable, today it is the territory whose shreds slowly rot across the extent of the map.(1)**

What is lost in this transition, Baudrillard tells us, is "all of metaphysics" (2). The distinction between being and appearances, "the real and the concept"—as if the former were the foundation and the latter its simple reflection—can no longer be maintained. So much for our faith in substrata, in the capacity of the concept to be "the mirror of nature." This mirror has cracked. No, not cracked. The image of a cracked mirror leaves too much intact. So not cracked but ***shat-t-t-t-tered***. It is now the appearances, the models themselves, that produce the real. "The real is produced," Baudrillard says, "from miniaturized cells, matrices, and memory banks, models of control—and it can be produced an infinite number of times from these." This "real" is no longer real but **hyper**real—more real than real. The era of simulation, Baudrillard notes, "is inaugurated by a liquidation of all referentials." Everything has slid into what Plato called a third order of representation . . . and it has, like the Energizer bunny, kept going and going and going. . . . Everything is always already a simulation now. Baudrillard says that "God himself, can be simulated, that is to say, can be reduced to the signs that constitute faith" (3).

> *RL is just one more window . . . and it's usually not my best one.*
> (Qtd. in Turkle 13)

This is Baudrillard's redistinction:

Representation stems from the principle of the equivalence of the sign and the real.

Simulation, on the contrary . . . stems from the radical negation of the sign as value, from the sign as the reversion and death sentence of every reference. (4)

On the one hand: What we hold against the MOOer is that her virtual constructions fail to maintain a clear

On the other hand: MOOspace is celebrated for **protecting** the reality principle by operating as reality's *other*. It, like Disneyland, figures as a "deterrence machine set up in order to rejuvenate the fiction of the real in the opposite camp" (13). Baudrillard notes that the world of Disneyland "wants to be childish in order

distinction between the real and the illusion—more to the point: Virtual Reality sucks Real Life into the black hole of simulation. It scores a point for imagology. And we-ideologues cannot have that. Enter the net-police—protectors of RL citizens, RL values, the reality principle itself. This border patrol takes very seriously the threat cyberspace poses to RL. The net-police attempt to save us from the downward spiral of simulation, to absorb its effects by interpreting it as ***false*** representation.

to make us believe that the adults are elsewhere, in the 'real' world, and in order to conceal the fact that true childishness is everywhere" (13). Those who renounce MUDding and MOOing as *only gaming* are actually building an argument for "reality" and the reality principle—by pronouncing VR a *pure fiction,* they indicate that something more authentic, more original, exists *elsewhere.* They celebrate cyberspace inasmuch as it makes "reality" seem all the more real and human beings seem all the more powerful.

Everyone knows that *Real* Life, *real* education, *real* community, etc., happens elsewhere, not in cyburbia but in the ***Real*** World (but not in MTV's *Real World,* which is only one more simulation). Enter, then, the humanist proponents of VR, such as Jaron Lanier, who are overanxious to confine the celebration of this technology within the bounds of human production and imagination. He wants to call Virtual Reality ***Intentional*** Reality. Why? Because VR, according to Lanier, is simply one of the many wonders created and controlled by ***his majesty the ego***. It is not a threat to Real Life at all, he suggests; on the contrary, "ultimately, everything is done by people and technology is just a *little game we play*" (Lanier 49; emphasis added). Cyberspace is not a threat to RL because it could never, contra Martin Heidegger, "slip from human control." (Heidegger 9)

These two perspectives, opposite as they appear at first, end up having ***everything*** in common. They sprout from the same serious metaphysic, operate within the same circle of constraint. No excess here; no questioning of the age-old distinction between "fantasy" and "reality." Neither position could even be thought except through a fundamental *forgetting* of that which Derrida asks us to re-remember:

> **[T]he conscious suspension of play . . . was itself a phase of play; . . . play includes the work of meaning or the meaning of work, and**

> **includes them not in terms of knowledge, but in terms of inscription: meaning is a function of play, is inscribed in a certain place in the configuration of a meaningless play.** ("From Restricted" 260)

So . . . there it is*[n't]*. We ask, then, with Turkle, "How do we keep a sense that there is a reality distinct from simulation?" And further: "Would that sense itself be an illusion?"(73). Is there a way to draw a distinction between VR and RL that doesn't allow the one to constantly bleed into the other? And if not, how do we distinguish between idle chat and productive dialogue? Playing and working? Tripping and Being-authentically? Here's where ***things get wild.***

MUDspace foregrounds issues of identity and representation in an in-your-face sort of way. Your first task as a new MUD resident is to name and describe whichever self it is that you intend to perform in this domain. You construct a *persona,* a character to inhabit in the MUD that may have nothing to do with any persona you inhabit in RL (which is not to say that it has nothing to do with "*you*"). These days, anything goes. Though a few years ago online cross-dressing could spark a scandal, it now

"Whereas representation attempts to absorb simulation by interpreting it as false representation, simulation envelops the whole edifice of representation itself as a simulacrum." (Baudrillard 6)

The Mr. Bungle affair. "A *Rape* in Cyberspace"?[4] Now, there's a good example of simulation/ reality seeping into reality/simulation. It certainly sparked some serious scrambling among the netizens on LambdaMOO. Their mission: to get some kind of grip on where MOOspace rubs itself against "reality" and where techno-bodies join their fleshly counterparts. What was suddenly, in a few horrible moments, crystal clear was that ***nothing at all*** was clear. Mr. Bungle was "playing" a "Bisquick-faced clown" in a "fake reality" (there's an oxymoron) that consisted of electromagnetic signals and a sophisticated database program with subprograms but with ***no*** organic beings, ***no*** material substances or referentials. This sounds safe enough; no "reality" here; it's pure simulation. And yet, what's so striking is that Mr. Bungle's simulation, his "play," refused to confine itself to the hard drive on which it "took place." Indeed, it appears to have seeped from the VR of LambdaMoo's living room into the RL homes and offices of several LambdaMOO "players" who were jacked in around the world, from New York City to Sidney, Australia.

Let's be as clear as possible about this: the textual cues that scrolled across the players' screens evoked a *sense,* an *image* of violation, which then emotionally affected the organic bodies reading the textual cues. Though we'd like to say this was ***no***/thing, it was clearly ***some***/ thing. A VR to RL seepage manifest itself in the RL emotional trauma expressed by some of Mr. Bungle's onscreen victims. This, as Heidegger liked to say, was ***not nothing***. VR had sprung an undeniable leak. And this was only one of many. A critical overflow now haunts the fluidifying border zone between these two worlds.

And in case you're thinking, "This is silly—they should have logged off"—*think again.* If you're still assuming it's as simple as logging off or just saying "no," you haven't been listening carefully enough; or, perhaps you haven't been listening with an ear that can hear this. It'll be necessary that you tune in with your Other ear, your *third* one, if you have any hope of hearing what wants to be said: ***This technology may not come with an off switch.*** It's looking as if its primary mode of being is set to "ON." It's operating, extending itself beyond itself, continuously affecting what we call "reality," ***whether you're jacked in or not.*** These worlds are *on each other*; they are under each other's influence. There's no prophylactic—no

***The reality engine generates virtual worlds even when we think they are real. Who can be sure whether or not she is wearing the goggles of a hidden simulator? Cultural programs, after all, code perception and cognition every bit as much as computer programs.* (T&S, *"Virtuality"* 2)**

barely rates a raised eyebrow. In fact, we might call it gender *sliding* since gender is not an either/or choice: one might gender oneself neuter, plural, splat, or "Spivak" (which is something like hermaphroditic), for instance.

And *gender* sliding is not the only border running activity in cyburbia. The human/nonhuman and human/machine boundaries also fluidify in VR. Entire MU communities are devoted to furry-animal and toon personae. And softbots wander MUD-space *as* human (and, presumably, as furry) characters. Here, men perform women,[5] humans perform furries and toons, and bots perform humans! But if you're thinking this is "only make-believe," listen up: "Playing" with identity in VR challenges what goes

word condom—thick enough to prevent leakage, to stop the co-mingling of VR and RL. And there is no longer any reason to suppose that there is a "natural world," a drug-free zone, into which we might retreat, *if there ever was.*

[I]n the past psychoanalysis would not have been what it was (no more so than many other things) if electronic mail, for example, had existed. (Derrida *"Mal d' Archive"*)

by the name identity in RL. Sandy Stone notes that, "many of the prenet assumptions about the nature of identity ha[ve] quietly vanished" (81). Identity *construction* inevitably calls attention to that which would claim to be **un**constructed. Is there really a stable "I" behind all the masks? Or is it, as Stone says, "personae *all the way down*"? (81).

Hey,

are you

buckled in?

Cyborg Crossings—@gender me fluid:

"Drugs," as Ronell notes, is an "impacted signifier," standing in for a "number of felt intrusions." Drugs, for this reason among others, "resist conceptual arrest" (*Crack Wars* 51). Indeed, they circulate everywhere, promise everything, but they remain aloof from the confines of any one disciplinary boundary. And though they are "dealt with" on almost all fronts, these dealings are all too frequently prohibitive, all too frequently conducted in the name of *authentic* or *natural* Being . . . as opposed, that is, to artificial Being, Being-*under-the influence,* Being-*on-something.* We really ought to stop ignoring what the war on "drugs" (and I include here literature, VR, and other "artificial" and/or "injectable" substances) has forgotten to take into account—that is, the "technological condition," as Derrida calls it:

I would suggest only that one consider the degree to which the literary object has itself been treated juridically as a drug. (Ronell, *Crack Wars* 55)

> **The natural, originary body does not exist: technology has not simply added itself, from outside or after the fact, as a foreign body.**

> **Certainly, this foreign or dangerous supplement is "originarily" at work and in place in the supposedly ideal interiority of the "body and soul." It is indeed at the heart of the heart**. ("Rhetoric of Drugs" 15)

Before we began to house electronic culture—*before* we re-cognized ourselves as cyborgs, jacked in, and moved to cyburbia—a technology of the human was always already circulating. All these conceptual border zones . . . authentic/inauthentic, human/machine, interior/exterior. They scan as linguistic guardrails for what is called "humanity," but they don't really exist; they have no essence or sub/stance; they're phantasmatic oppositions that desperately need to be re-imagined, especially since they're no less potent for their holographic "essence." They are, like the phallus that sired them, powerful but *empty*.

And let's get one thing (twistedly) straight: MOOspace hasn't managed to shake them. This new inflection of Being is not ideal. It would be silly to argue *for* VR *over* RL. MOOspace, after all, harbors all sorts of very familiar and not particularly savory desires—the desire to transcend the body, to master nature, to ascend to the level of a god (or at least a wiz[ard]), to mention only a few. This space, as Ronell notes, "is dependent on classical tropes of representation, imagination, the sovereign subject and negated otherness" ("Activist" 298). There's plenty of "gender trouble" in VR, lots of racial tensions and class struggles. No doubt about it, VR is the offspring of a "number of metaphysical cravings." But, on the other hand, and here's the point, "who isn't?" (299). What may be more important is that this technology is *also* busily challenging the assumptions upon which those cravings are based.

> ***Many bots roam MUDs. They log onto the games as though they were characters. . . . bots help with navigation, pass messages, and create a background atmosphere of animation in the MUD. . . . Characters played by people are sometimes mistaken for these little artificial intelligences. . . . And sometimes bots are mistaken for people.*** (Turkle 16)

Our machines, as Donna Haraway notes, have become "disturbingly lively" (194); and it ought not go unnoticed that *their* liveliness has dispersed *us*. Much to humanism's chagrin, Being goes rhizomatic in the cyburbs. It's beyond our control. Here, "the metaphysical subject," Ronell notes, "*is broken up* and displaced into routes of splintering disidentification" ("Activist" 299). Virtual subjectivity flickers at the intersection where technology corrupts humanist metaphysics. In fact, VR's fluidified subject *re*-articulates something Nietzsche noted way before any of us experienced our first MUD or data glove: that what humanism hopes is ***the*** subject is actually already a subject-as-multiplicity, a *community of the self.* And virtual worlds, to their credit, have had the guts to hail the "polyethoi" (as Victor J. Vitanza [184] calls it) *as such.*

In VR, the law of noncontradiction is often blatantly ignored. Coded boundaries that support notions of (gendered) power relations, identity politics, and human production, are unceremoniously transgressed. In cyberspace, we can be everywhere at once and take on any number of identities wherever we go. Indeed, "hierarchical dichotomies that have ordered discourse since Aristotle," Haraway says, are "techno-digested" (205) by an order of microelectronic simulation that makes interfaces (with the Other) *everywhere.*

> ***The cyborg, the multiple personality, the technosocial subject, Gibson's cyberspace cowboy all suggest a radial rewriting, in the techno-social space, . . . of the bounded individual as the standard social unit and validated social actuant.*** (Stone 43)

In "Cyborgs," one of *Imagologies'* minisections, Mark Taylor describes his struggle with diabetes. He notes that his life is sustained by a critical substitute—"synthetic insulin." (4). Here's Taylor:

> *Without the products of recombinant DNA, my so-called natural body would be dead. Three times a day I inject an "artificial" substance into my body. I suppose one could call this an addiction of sorts. The fate of being an insulin junkie is hardly an easy one. While I would never suggest that all drugs and addictions are life-sustaining, I am coming to suspect that the differences are not as great as those who have the health and wealth to be moralistic usually insist. As I slip the syringe into my leg, I realize that I share more with the addict in the ghetto than I ever dreamed possible. (4)*

Accompanying the recognition that we are cyborgs is a fundamental collapse of structures and an unnerving tolling of the knell *(un coup de glas).* It tolls, as Derrida notes, "the end of signification, of sense, and of the signifier" (*Glas* 31). Why? Because when we re-

Whether we like it or not, we *are* the illegitimate and so, as Haraway says, *unfaithful* offspring of war machines, human animals perpetually *jacked in* to technology's machinations. And our existence as such testifies to the breakdown of what Haraway calls "three crucial boundar[ies]," human versus animal, human-animal versus machine, and physical versus nonphysical. The

cognize ourselves as cyborgs, we lose the capacity to distinguish, finally, the interior from exterior, the me from the not-me, the "natural" from the synthetic. *And this is where things get funky.* The line between people and machines has gotten incredibly fuzzy. But new technologies are not responsible for *creating* this fuzziness; the line was always already a *ph*antasy, and that fact[oid] simply gets the spotlight in cyberspace. "There is, of course," as Stone notes, "**nothing fortuitous about these developments**" (44). The point here is neither to celebrate nor mourn this situation but only to note that "it is happening" . . .

whether we like it or not.

Without genetic engineering, I would probably be dead. I can continue to live only because I have become a cyborg. **(4)**

border wars fought to keep organism and machine separate were all about "production, reproduction, and imagination." But the cyborg takes "pleasure in the *confusion* of boundaries" (191).

Some Conclusionary [R]Amblings:

Cyberspace has problematized our specular image, spawned new forms of subjectivity, and extended the limits of Being into nonsubstantial spaces. The cyborg is the metaphysical/specular/*memorized* subject grazed by schizophrenia, broken up *and breaking up*—but this is not to say that it isn't producing, that there is no production in this "seduction." It only suggests that our notions of what it means to "produce" have been unexpectedly narrow. Cyborgs are not limited by the reflection they/we see in the mirror. If, under technology's influence, we are seduced to produce and commanded to master, we are also invited to let go, to unhinge, to *crack up.* This technology shoots the finger at metaphysics. What Lanier and company have missed is that this so-called *Intentional* Reality is perpetually *exceeding* our intentions and taking control of us in ways we have not yet begun to comprehend. Indeed, even as the subject's desire for mastery leads it to the cyburbs, cyburbia is busily splintering off the subject who would master. Disidentified, decentered,

[Some MOOers] experience their lives as a "cycling through" between the real world, RL, and a series of virtual worlds. I say a series because people are frequently connected to several MUDs at a time. (Turkle 12)

fluidified, fractalized—cyborgs are hardly what Descartes and Kant had in mind.

Taylor and Saarinen have suggested that the "cultural imperative is no longer: 'Get your shit together!' But: 'Let the shit hit the fan!'" ("Shifting Subjects" 2). Multiple personalities are *expected* in cyberspace—the *incapacity* to engage your **fractal interiorites** is a handicap. If you can't manage several different personas, connections, contexts at once, you might consider yourself *distributively challenged*. Like it or not, a propensity for rhizomatic zoomings characterizes the cyborg of the electronic age.

This obviously poses some problems for education and the pedagogical imperative. Any well-coded cyborg can perform the "unified subject" when that performance is required, when teachers, for instance, hail their students as what Stone calls a "BUGS—a Body Unit Grounded in a Self" (85). The role of a BUGS will, no doubt, remain in our repertoire for quite some time. But when that role gets performed as simply one among many, with no special privileging, what we call "learning" and "knowing" take a turn for which we have only recently—long after the fact—begun to flip our blinkers. Consider Turkle's description of one freshman's "study" habits:

> **In an MIT computer cluster at 2 A.M., an eighteen-year-old freshman sits at a networked machine and points to the four boxed-off areas on his vibrantly colored computer screen. "On this MUD I'm relaxing, shooting the breeze. On this other MUD I'm in a flame war. On this last one I'm into heavy sexual things. I'm traveling between the MUDs and a physics homework assignment due at 10 tomorrow morning." (13)**

This student is fulfilling the requirements demanded by a host of scattered subject positions, switching codes so quickly that Deleuze and Guattari would be proud. Note not only that so-called RL (the physics assignment) shows up ***as*** a window on his computer screen but also that it occupies only *one* window, that VR outnumbers it by a whopping three to one. And this scenario is rapidly becoming less the exception than the rule. So, as Turkle asks, "Where does real life end and a game begin?" "Is the real self always the one in the physical world?" And finally this: "As more and more real business gets done in cyberspace, could the real self be the one who functions best in that realm?" (241).

> ***[T]he boundary between science fiction and social reality is an optical illusion.*** (Haraway 191)

Don't even bother hitting the brakes. The turn into VR has already been taken. Or, to switch metaphors again, this pill has already been popped—we said yes before we got the chance to *just say no*. We're under the influence of a "hallucinogen" that is effecting a radical rupture in what we imagined were our limits, a radical refashioning of what pedagogy, knowledge, and "reality" *can be*. Retroactive prohibition won't help: resistance, as the Borg says, is futile. The best we can do at this point is scramble to catch up and, perhaps, tinker with dosage and mixture. This is risky business . . . we've already noted that this trip will cost us. And/but if we pay attention, there may also be a payoff—this space may well teach us a thing or two about the assumptions buttressing that which we've been trying to teach. And ***there's*** a good reason to

just

say

yes.

Notes

[1] Cf., Avital Ronell's "The Worst Neighborhoods of the Real."

[2] All references to Taylor and Saarinen (T&S) will indicate the separately numbered and titled sections of *Imagologies: Media Philosophy*.

[3] Certainly, Timothy Leary deserves more than a passing mention here as the cyberpunk who originally popularized the net surf as a high-tech acid trip. No doubt about it: Taylor and Saarinen owe one to Leary, particularly for illustrating the camouflaged connections between Haight-Ashbury and Silicon Valley. As T&S suggest, "The counter-culture's technophobia always harbored a technophilia that promised to transform the chemico-religious prosthesis into the electronic prosthesis" (9). Here, of course, we're suggesting that the notion of VR as a prosthesis doesn't exactly cut it. Unless, that is, we refine the connotations a bit. When Nietzsche recognized that there is nothing behind the mask, for instance, what goes by the name *mask* got a fashion update. Similarly, when we recognize that there's nothing on which to hook a prosthesis of any kind, we might consider that what we have here is something like "prostheses all the way down." Now, we're getting where we want to go.

[4] See Julian Dibble's "A Rape in Cyberspace" for a detailed account of this . . . "event."

[5] It's not at all unusual for men to gender themselves female in

MOOspace—those players gendered female get a lot more attention from other MOOers than those gendered male. Women also perform men, of course, and it's important to note that they often do that to avoid being *harassed* by other male-gendered players. Gender trouble is alive and well on the net. But the difference is that in this space, one performs whatever gender one wants to perform at the moment, and in the process, gender itself is sometimes stretched beyond recognition.

Works Cited

Baudrillard, Jean. *Simulacra and Simulation*. Trans. Sheila Faria Glaser. Ann Arbor: University of Michigan Press, 1994.

Derrida, Jacques. "From Restricted to General Economy: A Hegelianism without Reserve." In *Writing and Difference*. Trans. Alan Bass. Chicago: University of Chicago Press, 1978.

———. *Glas*. Trans. John P. Leavey, Jr. and Richard Rand. Lincoln: University of Nebraska Press, 1986.

———. "The Rhetoric of Drugs: An Interview." Trans. Michael Israel. *Differences: A Journal of Feminist Cultural Studies*. 5, no. 1 (1993): 1–25.

———. "Mal d'Archive: Un Impression Freudienne." Archive maintained by Peter Knapp at http://www.cas.usf.edu/journal/feb/arch.html.

Dibble, Julian. "A Rape In Cyberspace." *Village Voice*, December 21, 1993, 36–42.

Haraway, Donna. "A Manifesto for Cyborgs: Science, Technology, and Socialist Feminism in the 1980s." In *Feminism/Post-modernism*, ed. Linda J. Nicholson. New York: Routledge, 1990.

Heidegger, Martin. "The Question Concerning Technology." In *The Question Concerning Technology and Other Essays*. Trans. Albert Hofstadter. New York: Harper and Row, 1971.

Lanier, Jaron. "Life in the Data Cloud: Scratching Your Eyes Back In." Interviewed by J. P. Barlow. *Mondo 2000* (1990): 2–50.

Ronell, Avital. "Activist Supplement." In *Finitude's Score: Essays for the End of the Millennium*. Lincoln: University of Nebraska Press, 1994.

———. *Crack Wars: Literature, Addiction, Mania*. Lincoln: University of Nebraska Press, 1992.

———. "The Worst Neighborhoods of the Real." In *Finitude's Score: Essays for the End of the Millennium.* Lincoln: University of Nebraska Press, 1994.

Stone, Allucquere Rosanne (Sandy). *The War of Desire and Technology at the Close of the Mechanical Age.* Cambridge, Mass.: MIT Press, 1995.

Taylor, Mark, and Esa Saarinen. *Imagologies: Media Philosophy.* New York: Routledge, 1994.

Turkle, Sherry. *Life on the Screen: Identity in the Age of the Internet.* New York: Simon and Schuster, 1995.

Vitanza, Victor J. "Taking A-Count of a (Future-Anterior) History of Rhetoric *as* 'Libidinalized Marxism' (A PM Pastiche)." In *Writing Histories of Rhetoric.* Carbondale: Southern Illinois University Press, 1994.

13

Of MOOs, Folds, and Non-reactionary Virtual Communities

Victor J. Vitanza

Orientation

> Thought thinks its own history (the past), but in order to free itself from what it thinks (the present) and be able finally to "think otherwise" (the future).
>
> —Deleuze, *Foucault*

> The Baroque refers not to an essence but rather to an operative function, to a trait. It endlessly produces folds. It does not invent things: there are all kinds of folds coming from the East, Greek, Roman, Romanesque, Gothic, Classical folds. . . . Yet the Baroque trait twists and turns its folds, pushing them to infinity, fold over fold, one upon the other. The Baroque fold unfurls all the way to infinity.
>
> —Deleuze, *The Fold*

DisOrientation

While I and my kind were folding paper into airplanes and spitballs, other kids in the orient were being introduced to origami, the art of paper folding . . .

While my colleagues were demonstrating their abilities in class at unpacking a poem via formalist readings, showing a metaphor here, seven types of ambiguity there, I was cutting newspaper sheets into long strips, giving them a half fold, taping them together, and "cutting" them down the middle, showing students how a half fold could change the conditions of a material world . . .

A dividing practice (species-genus analytics, *diaeresis*) is a terrible way to unpack, unfold a paper. A terrible way to invite readers to project what will have been said and unsaid. It is a narrative strategy, a writing protocol, that determines the end. I . . . 'we' . . . my incompossible selphs[1] . . . will have to find a way to unfold and refold such a way of having been taught to write (fold, twist) a paper. To narrativize. And yet, it is important initially to mis/understand that unfolding is not the contrary of folding. Unfolding "here" will have been something else. Gilles Deleuze tells us: "Folding-unfolding no longer simply means tension-release, contraction-dilation [reactionary thinking], but enveloping-developing, involution-evolution" (*Fold* 9).

I guess, in this meditation, I will have to play with giving—"gifting"—everything a half fold, or in another phrase *a half twist.* I don't have an angle in or for this paper; I have a twist! And it . . . well, giving a rationalization . . . I can tell you it happened in this mannerist way: I said to one of my incompossible selphs, *You see, V, I would remove not only the center but also the edge. I would create a paper both with and without an edge, introducing the condition of the impossibility of cutting it into two. If a rapier were to be used, cutting the paper, it would bring into being both an infinity with an edge but more importantly an infinity, with which to play, without an edge.* And with this topos and atopos, I—one of my selphs—began work and play and then wPoLrAkY on this paper. And yet one more thing wants to be explained: While this is a cut-and-paste-and-cut paper, it is *not.* Go Figure. Just Figure.

At present, you—Dear Reader—should feel a bit twisted, wrenched. It will have been a necessity not only to give this section of the book a half twist, but "the reader" of this section a half twist.

This pulp version that would rather desire to become pixelated versions, virtual versions, is about folds, unfolding, and refolding and specifically in relation to MUDs and MOOs and WOOs and any other possible permutations and combinations—yet without limits . . . therefore, a SERIES . . . —that might be perpetually unfolded and refolded. (Sometimes traditional ways of unfolding a paper need, so as to realize desire, to be rudely interrupted! There is no necessary line of thinking, but moments of escape to refold elsewhere.) To repeat: Folds and refolds are not contraries. (The language "here" is tricky, for while it common sensically needs to suggest something Platonic-Cartesian, it uncontrarily desires to unfold something Leibnizian-Deleuzian. Something Baroque. And then, something cum Lyotardian. Something scandalous. Something all Rhetoric. We, in other folds, will be ever attempting to think the unthought. Therefore, we will be less concerned with what goes today under the rubric of "critical thinking"; we will be also concerned with the Other/other of sufficient reason. And yet, will have thought to move beyond the state of victimized/saintly "O/otherness." Inevitably then, we will be concerned with *Thinking?* What will have been called Thinking?

We will be concerned—given the proper wave of folding a paper—specifically with

- Leibniz-Deleuze's thinking about folds and the Baroque. The immediate purpose, as suggested, will be to rethink the architecture of MUDs/MOOs. To refold them into Baroque Houses. In dis/order to refold our thinking about virtual architecture, we will have to refold our notion of "archi" and of sufficient reason (continuity) and the possible (determination). We will have had to multiply our principles. Mutiplicity, however, will also be without sets and subsets. In other refoldings, there is no species in a genus, etc. There are radical singularities (everything is possible). From the Platonic-Aristotelian-Cartesian principle of sufficient reason and the possible, therefore, we will have had to refold our thinking to the . . .

- Leibnizean-Deleuzean principle of indiscernibles and the compossible (incompossible). And with them, we will have had to consider a theory of motives that is not based on necessity but also on "inclination" and "inflection." The *difference,* from a Leibnizean-Deleuzean view, is not a contrary; the difference between necessity but also inclination and inflection will have allowed us to understand a Baroque concept of justice. Why is this important? Because a house of BaroqueMOO that we would allegorize (our polis, our cybersocieties/cyberieties in the face of anxieties), that we would build and dwell in allegorically, virtually, will be inclined to recapitulate the house of all MOOs. And hence, justice will be continued in its reactionary form. The virtual effects the actual; the actual, the virtual. Leibniz-Deleuze say: "The world is a virtuality that is actualized monads or souls, but also a possibility that must be realized in matter or in bodies" (*Fold* 104). A case in point: In a Cartesian world, Sextus Tarquin in all of his versions (they are all one!) rapes Lucretia. In Leibnizean possible worlds, in our BaroqueMOO with a multiplicity of apartments, however, there is a multiplicity of Sextus Tarquins. The conditions for the compossibility of justice, therefore, have changed. My motive and inclination for introducing into this discussion . . .

- Leibniz's treatment of Sextus and the rape of Lucretia, therefore, is to reopen (refold) the Question of "A Rape in Cyberspace." The MOOhouses that we build, the possible/compossible (incompossible) cyberieties that we construct, I am inclined to think, must rethink (refold) the question of A Rape in Cyberspace. Refold the concept of justice. (I am, of course, referring to Julian Dibbell's now-canonical article of events at LambdaMOO.) Creating and Participating in Building a MOO is the establishing of a Virtual Society. Again, this is a creative act, and yet societies are founded on rape narratives. Recall un/just a bit of history: In book 1 of Livy's *History of Rome,* there are three *founding* events: The rape of the Vestal, which gives history/hystery the twins; the rape of the Sabine women, which brings forth the Roman people; and the rape of Lucretia, which brings forth the Republic. In article 1 (there must be other articles for other rooms!) of Dibbell's *The History of MOOs,* there is a founding rape and a cyberietal response based on certain "democratic" motives and inclinations. But, apparently, always already a reactionary judiciary response. And yet once ever again, a recapitulation of history! Must it be so?

In sum, then, my concerns are with discovering ways, waves of furls, via Leibnizean-Deleuzean notions of the fold and the Baroque, to rethink the question of MOO-*an*architechtural space, but also to think of folds and this space that these folds establish as a means of our (singularly and collectively) political resistance that would not be reactionary.[2] Hence, "we" will, as Deleuze himself has, think of folds as Foucault had. We will examine Deleuze's account of Foucault's third axis, moving from the axes of Knowledge and Power to that of *Thought.* Which means, like Foucault et al., "we" will have had to think the paradox of "The inside as an operation of the outside" (Deleuze, *Foucault* 97). And yet, "we" will do this *incipiently, imminently!* "Here" we will search for a way of rethinking, refolding, the Question of designing, refolding, a society/cyberiety, *not* without rape, for this—Leibniz's thinking—is no baroque *eu*topian thinking of cyberiety; here, we will also reconsider the Question of virtual Rape without reactionary thinking.

But, Dear Reader, this is not enough, never ever enough! For rape is not only in the world we live in, but is also the very foundation for all our thinking, or what stands for thinking. Reactionary Thinking. Knowing this, I will move on to a scandalous enterprise, restarted by Lyotard in his *Libidinal Economy.* Yes, Yes, Yes, the Great Ephemeral Skin! It's "our" half twist! (This is it; watch for it.) That enterprise will have been initially a taking up with the paradox of repetition and difference (the death principle supporting life), on the one hand, and a being taken away with radical singularities (no repetition, hence every event unique), on the other. And then, taken away, we would search for a third arm (see Stelarc). Which will be a half twist. The Great Ephemeral Skin! The place with no negatives. Except with the perpetual negating of negatives. Our one paracreative act. None other. A place denegated as in: Not a paper with no twist and therefore an edge; not a paper with a full twist and still an edge that cuts (rapes); but a paper with a half twist and no edge, no cut, but if a cut (rape), one that does not exclusively sever one thing from another. The half twist here, to be sure, must not recapitulate the founding principle as rape. The founding principle, at best (!), must be unfounded. "We" must admit, it is too late for MOOs as MOOs. (I, of course, casuisitically stretch my point.) They have been founded as such . . . on rape. We must, however, be concerned with recapitulation! And yet, as Nietzsche's Zarathustra warns, not concerned in order "to redeem those who lived in the past and to recreate all 'it was'" (251). But in promising you the absence of the "white terror of truth," I cannot promise you the absence, in a MOO, of "the red cruelty of singularities" (Lyotard, *Libidinal* 241). Instead, "we" promise you a Red far, far, far Left of being human. MOOs are no place for human beings, unless "human" is reconsidered etymologically as the full *tasting* of nature. (I am not alluding to furry muckers!) Setting aside all *nomoi* (laws, customs, consensus)! For a full tasting! Can you, Dear Reader, attempt such a tasting? (Beyond centermental furry muckers?) A moving from male/female, female/male, to third atopoi, heterotopias. If not, then, I leave you to work out your redemption and salvation via the road to "it was."

A MOO is a topos made up of virtual texts. It is metaphor based. Programmers work with metaphors. Give a MOO programmer a metaphor and s/he will attempt to simulate it. I will "gift" (without returns) un/just such a metaphor. I will toss one into the discussion. Nothing more; nothing less. The Great Ephemeral Skin as Möbius (half twist) strip. So as to forget MOOs in dis/order to un/think a Baroque Möbius strip MOO. WOO+. Or *WOOmb*. I will toss a metaphor into the discussion. A metaphor of bovine flesh that will have been all libidinal surface. Which will have been affirmatively deconstructed by way of all surfaces, in other words, without any height or depth, but with all superficial folds. So as to create the conditions for a new compossible cyberworld, what Deleuze would call "a reorientation of . . . entire thought and a new geography" (*Logic of Sense* 132). This notion of surface, however, is not "superficial" in any common sense of the word that would stand contrary to a view of "truth" in the lower depths.

To be sure, you, Dear Reader, will only feel cheated, feel frustrated. You, too, Programmer. And Why? Because not *eu*topian (i.e., totally without rape), and, therefore, disappointing (enraging?) leading some readers to continued reactionary thinking? No, this is not salvation history. (The agenda is a series of rebeginnings. The possibilities of a narrative with beginning-middle-end leads us only back to the white, epistemic violence, of which Lyotard speaks, while rebeginnings and more rebeginnings lead to the red cruelty of singularities.[3] The rebeginnings will require all of us who attend to the tossed metaphor to "*dance* on the feet of Chance" [*Thus Spoke* 278], and if need be—Oh, we fallen ones—to dance "with tears in [our] eyes" (220). The MOO will have become a pagan theater across a new geography. A *WOOmb*. Where all along the surface, without height and depth, the principle of individuation will not exist. All of this will require non-reactionary thinking.

To think means to experiment and to problematize. (Deleuze, *Foucault* 116)

Baroque House

With the above set of white and red promises and this subtitle (which is not a subordinate and which folds outside the standard protocol of academic discourse, yet stays within it more than enough, to con-fuse the differences), we might expect now to have a Baroque house described. Or the question answered, How is a MOO *like* a Baroque house or vice versa? To be sure, Leibniz-Deleuze supply us with a sketch (*Fold* 5). (And Lyotard, along with Leibniz-Deleuze, reminds us of the shortcomings of the principle of similitude.)

And yet, to get an understanding of what a Baroque House might possibly be, Leibniz-Deleuze do not just give a simple description without problematizing it. They give (gift) a description that is folded and then unfolded in later pages, only to refold it again much later in pages, overflowing out of its academic frame on the way to infinity. What they gift, also, are a set of counterprinciples (though not contrary) to the commonsensical principles of what thinking says a house is or can be. So we will have to start with these counterprinciples to get to some idea of what should be called Baroque *an*architecture, what a Baroque house or, better, BaroqueMOO might be. (A house can be a Klein Jar, which is all inside, with no outside, or neither! A MOO can be such a topological figure also.) And then Baroque justice in the chiaroscuro of the Question of Rape in Cyberspace. (I hear someone asking: *Aren't all MUDs/MOOs already Baroque virtual houses?* I would think not necessarily so. The medium of a MOO does not automatically create the conditions for the compossibility of a Baroque environment. And again: "Is there an example of [a] BaroqueMOO?" If so, I am not aware of one as I conceive of such an atopos. And yet, all MUDs/MOOs have the potentiality for becoming a BaroqueMUD/MOO. And then, "Is this not a contradiction?" Or as Leibniz would say, a vice-diction.)

It is tempting for me to describe particular MOO-experiences as simulated foldings, unfoldings, and refoldings, for example, as manifested metaphorically in terms of cascading waves of words being illuminated on a monitor; or in terms of a wavering "liquid identity" (Sandy Stone) of metamorphosizing from male to female to hermaphrodite to other animal, to becoming-animal; or in terms of rhizomatic space, etc. And while these metaphors (as in a catachresis) are helpful, they do not explain how fundamentally different the paraconcept of the Baroque is for thinking a

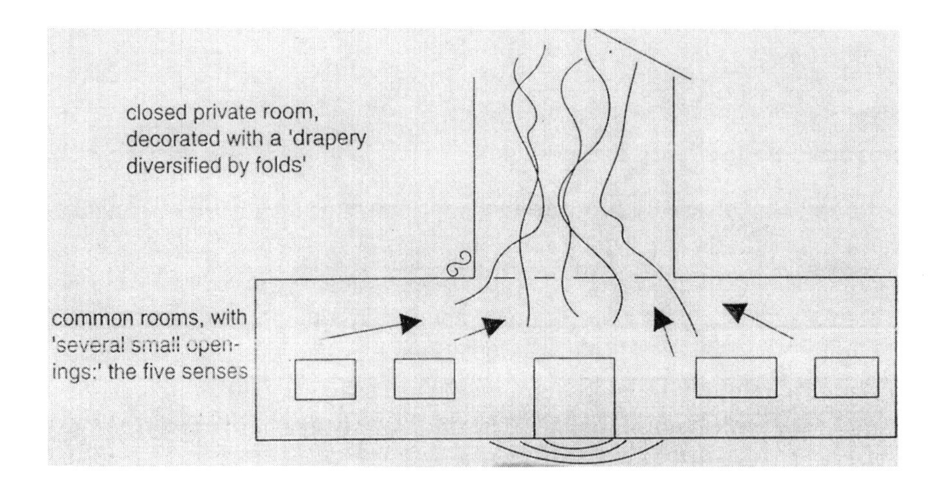

Fig. 12. The Baroque House (an allegory). (From Gilles Deleuze, *The Fold* [University of Minnesota Press, 1993]. Copyright © 1993 by the Regents of the University of Minnesota.)

BaroqueMOO. And so, I am going to unfold and refold the simulated conversation here of familiarities to some principles that Leibniz-Deleuze discuss, namely, that of *sufficient reason*/continuity and the *possible*/determination (Deleuze, *Fold* 41–58), on the one hand, and the *principle of indiscernibles* and *the compossible*/incompossible (59–75), on the other. (We are making our way to the red cruelty of radical singularities, but it's a way beyond this present discussion.) With Leibniz-Deleuze, we will take a look at the *multiplication of principles* (67) and the paraconcept of radical *singularities* (60–61). So my approach—Leibniz-Deleuze's approach—will be to call on abstract thinking to unfold State/Imperial thinking about thought so as to be able to refold it as Nomadic thought. As Leibniz-Deleuze say, it's a matter of refolding to remember what has been forgotten. And perpetually so, unto infinity. We will (ever here) suggest how to experiment and problematize (on our way, however, to affirmative forgetting).

The principle of sufficient reason (Plato-Descartes) is bound to—in bondage with—the three basic Master principles of Reason (Logic, the Great Zero), namely, identity, non-contradiction, and excluded middle. Either A is A or is not B or nothing else. A—in one steady-dominant fold—cannot be both A and B and C, etc. Sufficient reason could not allow for a refolding (problematizing) because, then, as Socrates-Plato were well aware, we would fall into infinite regress or, as Leibniz-Deleuze know, would slide laterally into ramifications of unfoldings and refoldings. The thinking that goes by sufficient reason, quite ironically, led to nihilism. For Leibniz-Deleuze, sufficiency (limit) was . . . *is* never, never, never enough. Hence, the principle of indiscernibles or Excess! (on their way to red cruelties). Therefore, the sophistical-casuistic thought presents itself: The Compossible (world), yet the incompossibilities (or monads). Let's take a closer look.

For Leibniz, Adam sinned, and yet he did not sin. In other words-folds, there are the singularities of Adam-A and Adam-B, who are equal: A = B. Is this a contradiction? Yes, if there is only one possible world, in which there cannot be such radical singularities. Or if there is only one possible way of Thinking about this problem, that Leibniz-Deleuze would (cannot not) problematize. Remember: Thinking is the issue here. For Leibniz, there are, in this unfolding-refolding example, at least, two ways of thinking about this problem and, hence, two possible worlds. (Actually and Virtually, there is an infinite number of possible ways. And those ways/waves are happening in this very text, Dear Reader, that you are reading, or attempting to read, and no doubt with great readerly trouble.) Deleuze explains, "Between the worlds there exists a relation other than one of contradiction. . . . It is a vice-diction, not a contradiction" (59). The contradiction can only exist in one possible world/monad. Here then, we have the Leibnizian paraconcept of Compossible and Incompossible. If the Possible world is limited by sufficient reason, the Compossible world is limitless (in excess) by indiscernibles, while each of the Incompossibles/ monads is limited. (The Possible is informed by a restricted economy; the Compossible, by a general economy, within which, again, there is an infinite number of monads, or radical singularities, with quasi-restricted economies. The best of all possible worlds is the white violence of Truth or imposed truth; the compossible world is the red cruelty.)

In Deleuzean terms, "[T]he world [is] an infinity of converging series, capable of being extended into each other, around unique points" (60). Allow me to illustrate, refolding back to an earlier attempt, with an example of a Möbius strip that when constructed from a narrow and long strip of blank paper has two parallel lines drawn on both sides of it and then, given *a half twist,* is/are taped together. If we start cutting with a pair of scissors on one line we will eventually see that we get to the other line (parallel lines do meet in this configuration/space) and will see that, when finished cutting, we have one strip (circle of paper), which is no longer a Möbius strip and another, which is a new Möbius strip. The two are joined, looped, or folded into each other. If we, then, cut the new Möbius strip, as we cut the previous one, we will get another non-Möbius strip (paper circle) and yet another new Möbius strip, and on ad infinitum. The Möbius strip is an infinity of converging series, capable of being extended into each other, around unique points. Möbius strip = The Compossible world; the paper circles/folds = monads, the Incompossibilities, each made up of singularities.

Now back to Deleuze, with my interpolation: "The result [in other words, of cutting the Möbius strip] is that another world appears *when the obtained series diverge in the neighborhood of singularities*" (*Fold* 60). He distinguishes:

> Compossibles can be called (1) the totality of converging and extensive series that constitute the world, (2) the totality of monads that convey the same world (Adam the sinner, Caesar the emperor, Christ the savior . . .). Incompossibles can be called (1) the series that diverge, and that from then on belong to two possible worlds, and (2) monads of which each expresses a world different from the other (Caesar the emperor and Adam the nonsinner). The eventual divergence of series is what allows for the definition of incompossibility or the relation of vice-diction. (60)

When God creates the world (keep in mind that Deleuze is describing Leibniz's view, not his view, which is by far more heretical than Leibniz's, and then there's Lyotard's yet to come), he sets forth "a series of . . . *pure emission of singularities:* to be the first man, to live in a garden of paradise, to have a wife created from one's own rib. And then a fourth: sinning" (60). Each of these singularities is held in a double (not contradictory, but vice-dictory) relationship. One line of the narrative is somewhat fixed like the paper circle cut from the Möbius strip; the other line is not fixed like the new Möbius strip. Their doubleness converges. (Incompossible converges, in other folds, with the Compossible.) Because they converge, a fifth singularity can appear: "resistance to temptation" (61). Deleuze continues:

> It is not simply that it contradicts the fourth, "sinning," such that a choice has to be made between the two. It is that the lines of prolongation that go from this fifth to the three others are not convergent . . . *do not pass through common values.* It is neither the same garden, nor the same primeval world, nor even the same gynegenesis. A bifurcation [a vice-diction] takes place that we at least take for granted, since reason escapes us. We are satisfied to know that one exists. It always suffices [principle of sufficient reason] to be able to say: that is what makes Adam the nonsinner to be supposed incompossible [principle

> of indiscernibles] with this world, since it implies a singularity that diverges from those of this world. (60).

And now from Adam(s) to his-their progeny. (Yes, let us not forget that this is all about not forgetting what has been forgotten and goes by the name of Thinking, that there are Adams that are nonsinners in other worlds/monads. If Compossibility, then, the characters or creatures that we meet, say, in a Borges tale should be open to our acceptance. What un/kind of MOO would Borges construct, what un/kind of cybereity and what un/kind of justice? The conditions for my rethinking this question are to be un/founded in Foucault's telling of Borges's tale of "order" according to "a certain Chinese Encyclopaedia" [*O* xv].)

The Baroque House That Is The Book

In figure 1, we have a sketch of The Baroque House (an allegory), which shows two floors. The top floor, Deleuze tells us, "has no windows. It is a dark room or chamber decorated only with a stretched canvas 'diversified by folds,' as if it were *a living dermis.* Placed on the opaque canvas, these folds, cords, or springs represent an innate form of knowledge, but when solicited by matter they move into action" (*Fold* 4; emphasis added). The bottom floor, which is connected to the upper, is "pierced with windows" and has "souls . . . sensitive, animal" (4) . This description is more complex, and we will return to it shortly. For now, however, I want to focus on another, later description of the House that is The Book. (Lest the reader forget, we are folding, unfolding, refolding our way/waves to thinking the thought of BaroqueMOO. And though this "I" teased you with Foucault's reference to Borges's tale, it will take a while to get there and beyond to the Ephermeral Skin, though as can be seen Leibniz-Deleuze speak of "a living dermis" when describing The Baroque House. We are far from, yet ever near the Ephemeral Skin.)

Deleuze, in passing, summarizes the "great Baroque staging [and narrative] at the end of [Leibniz's] *Théodicée*" (*Fold* 61), in which there

> is an architectural dream: an immense pyramid that has a summit but no base, and that is built from an infinity of apartments, of which each one makes up a world. It has a summit because there is a world that is the best of all worlds, and it lacks a base because the others are lost in the fog, and finally there remains no final one that can be called the worst. In every apartment a Sextus bears a number on his forehead. He

> mimes a sequence of his life or even his whole life, "as if in a theatrical staging," right next to a thick book. (61)

Deleuze continues:

> The number appears to refer to the page that tells the story of the life of this Sextus in greater detail, on a smaller scale, while the other pages probably tell of the other events of the world to which he belongs. Here is the Baroque combination of what we read and what we see. And, in the other apartments, we discover other Sextuses and other books. Leaving Jupiter's [or God's] abode [temple], one Sextus will go to Corinth and become a famous man, while another Sextus will go to Thrace and become king, instead of returning to Rome and raping Lucretia, as he does in the first apartment. All these singularities diverge from each other, and each converges with the first (the exit from the temple), only with values that differ from the others. All these Sextuses are possible, but they are part of incompossible worlds. (61–62)

Here, we have, unlike what we find in Susan Brownmiller's master metaphor for rape in *Against Our Will,* a view of a particular man who is a rapist but who also is not. For Brownmiller, all men are rapists. (Clearly, hers is a strategic essentialist and political position and understandably so. And yet, her strategy is reactionary.) Sextus in one incompossible world is a rapist; in another, he is not. And then, there are an infinite number of other versions. Remember: The possible narratives here, not informed by the principle of sufficient reason, are infinite. And a concept of justice? Is there one, or more, that goes with all this talk here? The simple answer is Yes. However, this, too, is a question, a thinking, that must be problematized.

A Deleuze-Leibniz-Borges Excursus

Therefore, let's problematize the condition for a question of justice by recalling Borges, just as Deleuze himself does in relation to what he says about Leibniz. Both Leibniz and Borges deal in bifurcations. For me, they themselves—the singularities, the incompossibility, of their visions—form *a bifurcation.* (We will have eventually taken a fork in the rhizomatic road.) Deleuze, at a tactical-pivotal point among others in his discussion, introduces Borges into his exposition of Leibniz's new logic. He says,

> Borges, one of Leibniz's disciples, invoked the Chinese philosopher-*architect* Ts'ui Pen, the inventor of the "garden with bifurcating

> paths," a baroque labyrinth whose infinite series converge or diverge, forming a webbing of time embracing all possibilities. (*Fold* 62)

More specifically now, Deleuze quotes Borges:

> Fang, for example, keeps a secret; a stranger knocks at his door; Fang decides to kill him. Naturally, several outcomes are possible: Fang can kill the intruder; the intruder can kill Fang; both of them can escape from their peril; both can die, etc. In Ts'ui Pen's work, all outcomes are produced, each being the point of departure for other bifurcations. (62)

The implicit question in Deleuze's discussion of Borges is why select Ts'ui Pen, instead of Leibniz as a primary example of justice/justification. Deleuze's answer: "Borges . . . wanted . . . to have God pass into existence all incompossible worlds at once instead of choosing one of them, the best" (62).

Yes, this traditional thinking of the *best of all possible* is the answer to the question of justice that must be further problematized. Deleuze, of course, explains that "what especially impedes God from making all possibles—even incompossibles—exist is that this would then be a mendacious God, a trickster God, a deceiving God. . . . Leibniz, who strongly distrusts the Cartesian argument of the nonmalevolent God, gives him a new basis at the level of incompossibility: God plays tricks, but he also furnishes the rules of the game" (*Fold* 62–63). At this point, Leibniz, as Deleuze renders him, must unfold and refold his arguments for the defense or the justification of the way/waves of God to all men! (The foldings and pleats of the justification are intricate, and they do not advance this present discussion.) The important thing to remember, however, is that God chooses not only the best of all possible worlds but also "chooses the best allotment of singularities in possible individuals." Deleuze continues: "Hence we have rules of the world's composition in a compossible architectonic totality" (66).

What this excursus is about to tell us, un/just as Deleuze's on Borges leads us to, is that Leibniz is perhaps situated somewhere *between* a beneficient God (Cartesian) *and* a trickster God or no God at all (Nietzsche and Mallarmé). Nihilism, as Deleuze recounts, presents itself, incipiently in Leibniz. And so, Deleuze asks: "But what happened in this long history of 'nihilism,' before the world lost its principles [of sufficient reason]?" (*Fold* 67). Or sufficient justice? His tentative answer:

> At a point close to us human Reason had to collapse, like the Kantian refuge, the last refuge of principles. It falls victim to "neurosis." But still, before, a psychotic episode was necessary. A crisis and collapse of all theological Reason had to take place. That is where the Baroque assumes its position: Is there some way of saving the theological ideal at a moment when it is being contested on all sides . . . ? The Baroque solution is the following: we shall multiply principles—we can always slip a new one out from under our cuffs—and in this way we will change their use. . . . Principles . . . will be put to a reflective use. A case being given, we shall invent its principles. It is a transformation from Law to universal Jurisprudence. (67)

Hence, the introduction of "the Leibnizian game" that "is first of all a proliferation of principles: play is executed through excess and not a lack of principles; the game is that of . . . inventing principles" (67–68). (Of course, we find this today in a variety of rhetorical and sophistical venues from Chaim Perelman's *The New Rhetoric* to Jean-François Lyotard's *Just Gaming* and *The Differend.* And find this thinking in a variety of virtual venuses from Allucquère Rosanne Stone, *The War of Desire and Technology at the Close of the Mechanical Age,* to Sherry Turkle, *Life on the Screen.*) The *best* that Leibniz can do at the time is to invent multiple Adams and Sextuses. In a typical atypical characterization, Deleuze writes: "We clearly witness a schizophrenic reconstruction: God's attorney convenes characters who reconstitute the world *with their inner, so-called autoplastic modifications.* Such are the monads, or Leibniz's Selves, automata" (68). But here's the clincher, as Deleuze sees it: "[H]uman liberty is not itself safeguarded inasmuch as it has to be practiced in this existing world. In human eyes it does not suffice that [Sextus] may not [rape] in another world, if he is certainly [raping] in this world" (*Fold* 69).

> They say he raped them that night. They say he did it with a cunning little doll, fashioned in their image and imbued with the power to make them do whatever he desired. . . . And though I wasn't there that night, I think I can assure you that what they say is true, because it all happened right in the living room—right there amid the well-stocked bookcases and the sofas and the fireplace—of a house I've come to think of as my second home.
>
> —Dr. Bombay (Julian Dibbell), "A Rape in Cyberspace"

The Question of Justice

We inhabit a world, in which rape is an everyday occurrence, and mostly suffered by women. We know that many rapes go unreported, and we know that they are an everyday occurrence in the world's prison systems. We for the most part think of rape as a physical, not a figurative, act. And yet, that view is changing; *the difference between the actual and the virtual is disappearing.* (Is the Libidinal Band, the Möbius Band, speeding up, taking us from the world of semiotic signs to the whirl of tensor signs, that is, the world in which differences collapse into radical singularities?) This difference between physical and figurative, however, is still a hotly debated topic. But the fact that it is debated gives support to the notion that the difference *may be* disappearing and that the discussion (deliberation) itself ironically speeds up the Möbius band, which in turn collapses differences even further.

In the now-canonized article "A Rape in Cyberspace," Dibbell—or perhaps I should say, Dr. Bombay, Dibbell's persona-avatar—examines the issue of difference in terms of judiciary discourse, jurisprudence.[4] Allegedly a rape occurs in cyberspace, and the argument, as usual, centers in great part on whether or not the act constituted a rape. As presented by Bombay, the "facts" of the case are that on "a Monday night in March . . . in the living room [at LambdaMOO, in a mainframe or server] at or about 10 P.M. Pacific Standard Time . . . legba, a Haitian trickster spirit of indeterminate gender," coded and being typed in by a person in Seattle, and then "Starsinger, . . . a nondescript female character" (239), in Haverford, Pennsylvania, were both raped by Mr. Bungle, a puppet, who was manipulated with code by "a young man logging in to the MOO from a New York University computer" (241). There were a number of "witnesses" in the living room, which was modeled virtually after Pavel Curtis's own living room in real life. There was a furor over the virtual act. Some—including legba, who made a formal motion, and Starsinger, Bakunin, and even SamIAm (the "Australian Deleuzean"!) seconded the motion—wanted "to toad" Mr. Bungle, that is, virtually to execute his being erased from the memory of the MOO (245). Soon, about fifty members of the virtual community agreed to the judgment. But as Dr. Bombay tells us: "There was one small but stubborn obstacle. . . , the New Direction" (245), which was a formal document on matters of policy that had been developed by Pavel Curtis (the archwizard, Haakon, and "principle architect" of LambdaMOO). The policy stated that "the wizards . . . were pure technicians. . . . they would make no decisions affecting the social life of the MOO, but only implement whatever decisions the community as a whole directed them to" (246). In other words, the members of the MOO were given the right of "*inventing* [their] own self-governance from scratch" (246; emphasis added). And yet, there was no legislation (*nomos*—customs, rules, laws) for handling The Affair Bungle that—let us not forget—took place in cyberspace.

What follows from that moment on in Dr. Bombay's report is a fall into jurisprudence and what emerges are different groups formed by a mutual acceptance of a variety of warrants. The language game that the community generally agreed to, however, was basic, informal argumentation. There were the "Parliamentarian legalist types [who] argued that unfortunately Bungle could not legitimately be toaded at all, since there were not explicit MOO rules against rape, or against just about anything else" (246–47). Next were those composing a "royalist streak" who wanted to do away with "this New Direction silliness" and to recall "the wizardocracy . . . to the position of swift and decisive leadership their player class was born to" (247). Then there were "the technolibertarians" and the "resident anarchists" (247–48). There was a call for a meeting on the third day so as to settle the case, but after a long meeting nothing was settled (250). Mr. Bungle shows up. Instead of his appearance acting as a catalyst to fire up the group, it only emphasizes their apparently hopeless situation.

Instead of ending in a state of entropy, something does happen. Present at the meetings were the wizards, one by the name of Joe Feedback, "who'd sat brooding on the sidelines all evening" (252), trying to decide whether he was going to do something about the case. He finally decides, on his own, to toad Bungle (253). We are told:

> They say that LambdaMOO has never been the same since Mr. Bungle's toading. They say as well that nothing's really changed. And though it skirts the fuzziest of dream-logics to say that both these statements are true, the MOO is just the sort of fuzzy, dreamlike place in which such contradictions thrive. (253).

The MOO here is described as Baroque, as having a *vice-diction*. And the MOO is not the same—and virtually was not the same since the origination of the problem by the act of cyberrape—for someone has virtually acted out the full narrative of crime and punishment.

Haakon (Curtis, archwizard), having been away from LambdaMOO, returns to see the "wreckage strewn across the tiny universe" and consequently adds a statement to New Directions saying that "he would build into the database a system of petitions and ballots whereby anyone could put to popular vote any social scheme requiring wizardly powers for its implementation, with the results of the vote to be binding on the wizards" (253). LambdaMOO was now a pure democracy. And so all's well that ends well. Not so! For a few days after the toading, "a strange new character named Dr. Jest" arrives. There is every reason to believe that this new character is "Mr. Bungle . . . risen from the grave" (254). Toading was not fatal because all that is necessary for the "dead" to rise again is to get a new account from another provider and then return to LambaMOO under a different name. The return of the repressed is automatic in one form or another! And to this day Mr. Bungle/Dr. Jest sleeps in his room at LambdaMOO. Waiting to awake.

Lest we think that some good did come from the event in terms of the revision of the New Direction, which established a pure democracy, we need constantly to remind ourselves of the larger event and who constructed this society. Though it was put into motion by the programming of the archwizard, Haakon, the society and its anxieties were all established by those who composed the cybersociety. The subtitle of Dibbell's article "A Rape in Cyberspace" reads "How an Evil Clown, a Haitian Trickster Spirit, Two Wizards, and a Cast of Dozens Turned a Database into a Society." And again, How? What Dr. Bombay-Dibbell make clear—though subtly—is that the How of the event is the "urBungle" (254). The bungling of the whole matter, which perhaps could only be bungled, is *the founding* event for LambdaMOO. First the cyberrape (crime), the attempt at justice (deliberation), the toading (the punishment determined and executed by one person), and then the regeneration of the Bungle-like character. And let us not forget the reactionary form that the punishment took and how it was unwittingly engineered, because of the media (software, hardware, wetware), to repeat itself perpetually. Like Adam, like Sextus, like Bungle, like Jest, etc. (Let us not forget that LambdaMOO was the original of its kind of experimental MOO, or world.) It can certainly be argued that the decree added to the New Direction should or would solve any future problems in terms of this narrative repeating itself, but the founding event can never be circumvented. Nor can anyone ever stop a wizard or the archwizard, as constituted wizards, from repeating the orginal act of punishment. Also having access to an *@gag* or *@boot* command, only deflects the problem, which is always already there waiting to happen. And of course, these commands can be used indiscriminately. Moreover, arriving at consensus itself *(homologia)* is predetermined by the urBungle, un/just as a question determines what its answer could be. Instead of one acting out the punishment, it's the many in the name of one acting. Lyotard so describes such deliberative (empowering?) Thinking (and I interpolate):

> *Thought* begins with the possible. This is why the *logos* begins with the *politeia* and the market. It is as the voice of writing [typing on the screen], the production of [semiotic] signs with a view to exchange,

monopolized almost all the libido of the citizen-merchant bodies. But I am not saying that the body that speaks, writes and thinks [types], does not enjoy. . . , rather that its charge instead of taking place in singular intensities, *comes to be folded back* [reterritorialized, returned] not only onto the need of the market and the city, but *onto the zero* [the lack, the Negative, the principle of Exclusion in the name of Justice] where both are centred, on the zero of money [virtual capital] and discourse. Nihilism . . . brings the empty *méson* [virtual community as political sphere] round which the members of the *koinonia* gravitate. (162–63; emphasis mine)

I would continue describing and summating, for the ur-narrative of LambdaMOO does repeat itself in terms of "it was." So many commonplace questions are left standing, and yet: Look at the questions! How they tantalize! If "we" refuse them, "we" are (will be) accused of being neoconservatives. And yet, "we" must refuse them. If we don't refuse them, they only fold us back onto the Zero. Let us recall Zarathustra's words:

> Alas, every prisoner [of a past misdeed] becomes a fool; and the imprisoned will redeems himself [and herself] foolishly. That time does not run backwards, that is his [her] wrath; "that which was" is the name of the stone he [she] cannot move. And so, he [she] moves stones out of wrath and displeasure, and he [she] wreaks revenge on whatever does not feel wrath and displeasure as he [she] does. Thus the will, the liberator, took to hurting; and on all who can suffer he [she] wreaks revenge for his [her] inability to go backwards. This, indeed this alone, is what *revenge* is: the will's ill will against time and its "it was." (251–52)

Charles Stivale has given an excellent account of the continuum of revenge and counter-revenge at LambdaMOO and how it all spills over into elsewhere. The history of this event of Lambda, as in the sign of Alpha/Omega, is *all human, too pathetically human,* without a taste for life/*Eros.* Stivale renders it as "it was."

As it was in the beginning, so it was at LambdaMOO. Let us discontinue. Let us go elsewhere. Affirmatively forget. Far far from here. Let us . . . refold . . . elliptically to . . .

> Consider[ing] the cattle, grazing as they pass you by: they do not know what is meant by yesterday or today, they leap about, eat, rest, digest, leap about again, and so from morn till night and from day to day,

> fettered to the moment and its pleasure or displeasure, and thus neither melancholy nor bored. This is a hard sight for man to see; for, though he thinks himself better than the animals because he is human, he cannot help envying them their happiness—what they have, a life neither bored nor painful, is precisely what he wants, yet he cannot have it because he refuses to be like an animal. (Nietzsche, "Uses and Disadvantages" 60–61)

. . . And so it was in those days that Cow did come forth, having forsaken

> the peace of death in return for the promise of renewed glory . . . and the Dancers of the Cow were plentiful, fulfilling the ancient rites with their purity and self-cleansing. The Chant did spring forth onto the Land. . . . The wind did carry the name of Cow to all who listened. "May the Promise of Bovine Restfulness be with us all. Let Humanity know that Cow has returned." Thus did the cry ring out—over hill, over dale did the name of Cow extend, until the Universe chanted in rhythm with the Dancers. The Holy Circle did expand, yea, unto a tumultuous whirlpool of raw emotion and festivity.
>
> . . . And Cow did proclaim with all heartfelt sincerity to those assembled, "Moo." The four winds blew across the world, taking Cow's wisdom wherever they went. (Akira)

A CAESURA (A Radical Interruption): On Our Way from Beyond BaroqueMOO to LibidinalMOO, WOOmb

And so we have arrived at the moment of What? Bovinity? Yes. And yet, No! For we will not bring the cows home; we will go home with the cows. We, the unhomely ones! What does *go home* mean? It means a folding of ourselves into new *ethoi* and then a new geography.

New Ethoi: In part, the etymology of the word *ethos* is *éthea,* which is the name of the place where animals belong and which is the name of the places/topoi where the barbarians, the non-Greeks go. Where the uncivilized dwell. (See Scott 142–47.) Going home means, therefore, we will leave the previously assured protection of the Archi-Wizards simulated living room for . . . Do you think, Dear Reader that I am merely suggesting that we fold from the Civilized Greek to the Non-Greek! Oh, NO! . . . We will leave the previously assured protection of the *Archi-* for the perpetual terror of singularities of the *anarchitechture* (*sic*) of "a Greek labyrinth which is a single straight line" (see Borges, *Ficciones* 69–70; cf. Deleuze, *Logic and Sense* 62). We will be nomadic drifters in the non-Greek space left in the midst of Greek Rationality. Under the thin crust of their foundations. In their methods of deliberation, etc. that only loop back to devour them. We will be not One, nor Two, but a billion and one.

A New Geography: And so I have suggested that we must change our *archi*tec(h)ture and its overall geography. Ha! But such a change is not social reengineering as it is commonly offered. I go too far. The only deliverance I can suggest to Apollonian programers and diggers is self-overcoming.[5] All things cows! Nothing cowers! There's no way to end or rebegin this meditation so that everyone will have been happy. And I can't be concerned with such silliness. While Borges's narrator awaits me in the straight and narrow, I just want to MOO. Not cower. MOO. MOO. MOO.

A new geography will be without heights and depths. There will be no heights of Platonic idealism (no Ideal forms) and no depths of Freudian, Marxist, and Nietzschean suspicion (no unconscious, ideology) to be colonized. Deleuze explains: "This is a reorientation of all thought and of what it means to think: *there is no longer depth or height.* The Cynical and Stoic sneers against Plato are many. It is always a matter of unseating the Ideas, of showing that the incorporeal is not high above . . . but is rather at the surface, that it is not the highest cause but the superficial effect par excellence, and that it is not Essence but event" (*Logic of Sense* 130). Deleuze points out more so *how* the realm of the depth ("subversion") "acts in an original [i.e., inventional] way, by means of its power to organize surfaces and to envelop itself within surfaces. This pulsation sometimes acts through the formation of a minimum amount of surface, for a maximum amount of matter" (124). What Deleuze gives (gifts) is an account of a theory/spectacle of how scarcity, lack, and depth are created. And more importantly, how we are manipulated by loss of access to (all libidinal) surface. Hence, while Platonism and its heritage denies the value of all surface and depth, a hermeneutics of suspicion—in promising to demystify the surface and make manifest the depths—reshapes the "truth" and controls the access to its presentation on the surface.[6] Where we would live. Everything is permitted on the surface. There are no rules, laws, truths. Everything mixes with everything. There is no container to contain. There are only the red cruelties of radical singularities. Where, as Deleuze says, there is a "reorientation of the entire thought and a new geography" (132). In this place, we would be like the Stoics, establishing ourselves and wrapping ourselves "up with the surface, the curtain, the carpet, and the mantle. The double sense of the surface, the continuity of the reverse and right sides, replace height and depth. There [would be] nothing behind the curtain except unnameable mixtures, nothing above the carpet except the empty sky." We would be like "the animal which is on a level with the surface—a tick or louse" (133). We would be like Nietzsche's cows, grazing the surface.

An announcement: All those held captive by the depths, the netherworlds, all those who are the subterraneans . . . RETURN TO THE SURFACE! RECLAIM THE SURFACE. It is y/ours! So that you might *self-overcome.*

In the meanwhile a question: *Can self-overcoming be simulated anarchitechturally?* Yes. Yes. Yes. It is possible to have an antiOedipal, anti-Electra, MOO. *Can it be anaprogrammed?* MOO, MOO, MOO, when you, Dear Programmers and Diggers, *figure* it out/in, in/out! Collapse the two. Reach for a nonsynthesized third. All Surface. Unthink and dig topologically. Not topological butterflies, but Cows! Make all things Cow surface. Cow geography. How to do this? But that's not my problem! Accept this as my gift. To transfer. Transference. To passover. To *self-overcome.* Being human, all too human. Become Cow. Becoming-Cow. Just flatten out the space of MOO into all surface. All sur-face. Diggers! Dig to the surface! Dig the surface. Dig? Cows have four stomachs. Dig? (If Kiki Smith can take human organs and turn them into installation art, I can take animal organs and de-domesticate them.)

> Except we turn back and become as cows, we shall not enter the kingdom of heaven. For we ought to learn what would it profit a man if he gained the whole world and did not learn this one thing: chewing the cud! He would not get rid of his melancholy—his great melancholy; but today that is called *nausea.* Who today does not have his heart, mouth, and eyes full of nausea? You too! You too! But behold these cows!
>
> —Nietzsche, *Thus Spoke Zarathustra*

"Our" Metaphor of BaroqueMOO Folding into BovineMOO into WOOmb; Or, Opening the Libidinal Surface

Open[7] the horn-, bagpipe-like organ and spread out all its surfaces, though they balloon and roll together: Not only the ovaries with their germinal epithelium, peritoneal epithelium, corpora lutea (with its pronounced yellow color), and graafian follicles (which surround the surface), but open and spread, expose the uterine cornua, longitudinally cut and flatten out the fifteen-inch sections, as they taper gradually toward the free end—noticing how the free part of the horn folds at first downward, forward, and outward, and then folds backward and upward, forming a spiral band—then the cervical canal, the muscular coat, the mucous membranes, uterine cotyledons . . . all hundred, either irregularly scattered over the surface or arranged in rows of about a dozen each . . . as though your dressmaker's scissors were opening the several layers of a finely tailored coat . . . go on to the uterine glands, long and branched; to the cervix, pale, glandless, and with numerous folds, noticing all along the way and especially at the external uterine orifice the folds (plicae palmatae) forming rounded prominences arranged circularly, which project into the cavity of the vagina . . . go on opening and flattening the recto-genital pouch of peritoneum and the two canals of Gartner, which are about the diameter of a goose quill, finally passing the blade along the way posteriorly near the external urethral orifice . . . into the vulva, with its thick, multilayered labia, continuing into the suburethral diverticulum, the two glandulae vestibulares majores, with their two or three ducts, which open into a small pouch of the mucous membranes, making your way, flattening the surface . . . all the way . . . giving it, ever giving it, a half (not a full!) twist!

And this is not all, far from it: Connected are the mammary glands, the udders, normally two in number, but often with countless more; armed with scalpels and tweezers, dismantle and lay bare the four teats of each udder, searching for the single duct, which widens superiorly and opens freely into a roomy lactiferous sinus, aka the milk cistern; then . . . take note . . . and learn . . . from how the mucous membrane of the sinus forms numerous folds, which render the cavity multilocular. In the libidinal body of the dead cow, all is fold or can be refolded into fold, by flattening the surface and ever giving it, along the waves, a half twist. Another. Another. Another Udder. The cow body, organized—dis/organized?—to . . . Cult of the Dead Cow (cDc).

All bodily unity layed "phlat" . . . a new political economy, libidinal. All Baudrillardian surface. All is seductive, immanently so! All these opened, half-twisted surfaces, "zones," as Lyotard says,

> are [to be] joined end to end in a band which has no back to it, a Moebius band which interests us not because it is closed, but because it is one-sided, a Moebian [ephemeral] skin which, rather than being just smooth [as we might have been led to believe], is on the contrary (is this topologically possible?) covered with roughness, corners, creases, cavities which when it passes on the "first" turn will be cavities, but perhaps on the "second," lumps. (*Fold* 2–3)

Yes, the band must turn . . . *fold rapidly,* accelerating, ever, ever rapidly so that the angles/angels of geometry become not just Euclidean but elliptical, hyperbolic, etc. spaces. Wherein the Euclidean Synecdoche, container, becomes a Klein Jar. Turning and turning. Now taking some paraterms from Baudrillard, the language *(logos)* of space moves from Duality through Polarity to Digitality. From triangular Oedipus to digital Narcissus (154–73). To Multiplicity. Ha! Programmers, wizards, can you dig? And yet, let us reinsert Baudrillard: "Anatomy is not destiny, nor is politics: seduction is destiny" (180).

And now and now and now, let's drift toward Foucault's Borges, who/which brought forth Foucault's own protoMOObook. Seduction calls us to Foucault's Borges. The Ephemeral band speeds up some more, as our hearts speed onward. Speeding. Ever Speedier. Foucault . . . panting . . . says that his ProtoMOObook

> arose out of a passage in Borges, out of the laughter that shattered . . . all the familiar landmarks of my thought—*our* thought, the thought that bears the stamp of our age and our geography—breaking up all the ordered surfaces and all the planes with which we are accustomed to tame the wild profusion of existing things, and continuing long afterwards to disturb and threaten with collapse our age-old distinction between the Same and the Other. This passage quotes a "Certain Chinese encyclopaedia" in which it is written that "animals are divided into: *(a)* belonging to the Emperor, *(b)* embalmed, *(c)* tame, *(d)* sucking pigs, *(e)* sirens, *(f)* fabulous, *(g)* stray dogs, *(h)* included in the present classification, *(i)* frenzied [and the list goes absurdly on]." In the wonderment of this [ana]taxonomy, the thing we apprehend in one great leap, the thing that, by means of the fable, is demonstrated as the exotic [seductive] charm of another system of thought, is the limitation [negating principle] of our own, the stark impossibility of thinking *that* (xv)

. . . thinking that a MOO "really" is a cow! A cow, a MOO. And both, surfacing as the third libidinal-Möbian un/kinds.[8]

Notes

1. The I/We distinction in quotes here and throughout and their misuse in terms of standard grammar have become a common way of specifying the elasticity and multiplicity of the speaking/writing/thinking subject. In other words, there is more than one writer of this article, with each being a fiction. Cf. Turkle, "Who Am We?"

2. My point here is that the issue is not just a mere concern with the architecture of electronic discourse or MOOs, but an ethicopolitical concern. Archi-tecture must be changed to Anarchi-TechTure, which is a neologism for denegating the Archi and the Techne of a philosophical, closed, Euclidean space. This paper, then, concerns the ethics and politics of the MOOscape, and is so concerned that it would radically change the very conditions of the possibilities of Architecture in terms of Anarchitechture, nullifying the *Archi,* while refusing to take as a

means of change an ethics and politics that would be reactionary and thereby only repeat the conditions of the Archi. My take is to build on but to depart from the work of Badiou, Boundas, Benjamin, Cooper, Nunes. But more so, my take is to dissipate by way of Georges Bataille. My rethinking of architecture is in many ways similar to Bataille's. For an excellent discussion of Bataille and architecture, see Hollier.

3. As I state, I borrow these terms of "white" and "red" from Lyotard. This white/red distinction may appear to be a binary; it is not, for there is a second element not included here, that of the binary or polar opposites. The sequence with the missing term would read: (1) *White Truth* (Unity or what would go for truth in a dominant discourse) (2) *Polarity* as White and Black, Male and Female, etc. (two sides, with one privileged over the other, but with possible reversals) (3) and *Red Terror* (radical singularities without the anchor of a "real" or fictional unity or polarity). The position of White Truth, I would venture to say, is totally unacceptable by most of us. The position of Polarity, which we probably will not agree on, especially allows in a so-called democratic framework for reactionary states of thinking structured in the name of justice. It is the Third position that is not a classic or modern position that I have been exploring as an alternative to the 1 and 2.

4. For an excellent discussion of Dibbell's article see R. MacKinnon; and Dibbell's "My Dinner."

5. The phrase "Apollonian programmers" is offered as a healthy challenge to programmers. Normally and understandably, a reference to anything Apollonian is taken as part of the vocabulary of a Nietzschean Apollonian-Dionysian split or fusion, which is one context that I intend here. However, when I say "healthy challenge," I am also referring to the other conditions of Apollo, who, though often seen as the god of mastery and control, is also, like Dionysius, "a god," according to Perniola, "who aroused in his followers . . . a raving delirium." Apollo represented the "principle of enigma[, that is,] the coincidence of the rational and the irrational." Perniola continues to say: "The nature of Apollo can only be defined using an oxymoron: terrible serenity, nonparticipant possession, delirious moderation, *sobria ebrietas*" (15–16). Apollonian programmers model thyselphs after such an oxymoron (after a Möbius strip). And How? By Self-Overcoming! (See Nietzsche, *Genealogy of Morals* 3.27)

6. Paul Ricoeur named Freud, Marx, and Nietzsche the "hermeneuts of suspicion." I disagree with his characterization of Nietzsche. Deleuze also does: "Nietzsche was able to rediscover depth only after conquering the surfaces. But he did not remain at the surface, for the surface struck him as that which had to be assessed from the renewed perspective of an eye peering out from the depths. Nietzsche takes little interest in what happened after Plato, maintaining that it was necessarily the continuation of a long decadence. We have the impression, however, that there arises, in conformity to this method, a third image of philosophers. In relation to them, Nietzsche's pronouncement is particularly apt: how profound these Greeks were as a consequence of their being superficial!" (*Logic of Sense* 129).

7. This is my beginning of a pastiche of Lyotard's opening paragraphs in *Libidinal Economy.* It is an appropriation. It is the creation, in one sense, of a body without organs (Deleuze and Félix Guattari), a laying flat of the political body

(Lyotard), and a topological restructing of the body so that it will be all surface (Baudrillard).

8. An ending, if folded, is a rebeginning: Upon final folding of this paper, I continued—and still continue—working and playing on the theme of *an*architechture, with Haynes, Holmevik, and Kolko. See our "MOOs, Anarchitexture, Towards a New Threshold."

Works Cited

Akira. "Dance of the Cow." Cult of the Dead Cow. Online. 11 July 1996. Available: http://www.l0pht.com/cdc/cdc147.txt.

Badiou, Alain. "Gilles Deleuze, *The Fold: Leibniz and the Baroque.*" In *Gilles Deleuze and the Theater of Philosophy,* ed. Constantin V. Boundas and Dorothea Olkowski, 51–69. New York: Routledge, 1994.

Baudrillard, Jean. *Seduction.* Trans. Brian Singer. New York: St. Martin's, 1990.

Borges, Jorge Luis. *Ficciones.* Ed. Anthony Kerrigan. New York: Grove, 1962.

Boundas, Constantin V. "Deleuze: Serialization and Subject-Formation." In *Gilles Deleuze and the Theater of Philosophy,* ed. Constantin V. Boundas and Dorothea Olkowski, 99–116. New York: Routledge, 1994.

Benjamin, Andrew. "Time, Question, Fold." 27 April 1997. Available: http://www.basilisk.com/V/virtual_deleuze_fold_112.html.

Conley, Tom. "Translator's Foreword: A Plea for Leibniz." In Gilles Deleuze, *The Fold.* Minneapolis: University of Minnesota Press, 1993.

Cooper, Wes. "Wizards, Toads, and Ethics: Reflections of a MOO." 27 April 1997. Available: http://www.december.com/cmc/mag/1996/jan/cooper.html.

Deleuze, Gilles. *The Fold: Leibniz and the Baroque.* Trans. Tom Conley. Minneapolis: University of Minnesota Press, 1993.

———. *Foucault.* Trans. and ed. Seán Hand. Minneapolis: University of Minnesota Press, 1988.

———. *The Logic of Sense.* Trans. Mark Lester with Charles Stivale. Ed. Constantin V. Boundas. New York: Columbia University Press, 1990.

Dibbell, Julian. "A Rape in Cyberspace." In *CyberReader,* ed. Victor J. Vitanza, 448–65. Needham Heights, Mass.: Allyn & Bacon, 1996.

———. "My Dinner with Catherine MacKinnon." Available: http://www.levity.com/julian/mydinner.html.

Foucault, Michel. *The Order of Things.* New York: Vintage, 1973.

Haynes, Cynthia, Jan Rune Holmevik, Beth Kolko, and Victor J. Vitanza. "MOOs, Anarchitexture, Towards a New Threshold." Forthcoming in *The Emerging CyberCulture: Literacy, Paradigm, and Paradox.* Ed. Stephanie Gibson and Ollie Oviedo. Hampton Press.

Hollier, Denis. *Against Architecture: The Writings of Georges Bataille.* Trans. Betsy Wing. Cambridge, Mass.: The MIT Press, 1989.

Lyotard, Jean-François. *Libidinal Economy.* Trans. Iain Hamilton Grant. Bloomington: Indiana University Press, 1993.

MacKinnon, Richard. "Virtual Rape." *Journal of CMC* 2.4 (March 1997). Available: http:/jcmc.mscc.huji.ac.il/vol2/issue4/mackinnon.html.

Nietzsche, Friedrich. *On the Genealogy of Morals* and *Ecce Homo.* Trans. Walter Kaufmann and R. J. Holingdale. New York: Vintage, 1969.

———. "On the Uses and Disadvantages of History for Life." In *Untimely Medi-*

tations, Trans. R. J. Hollingdale, 57–123. Cambridge: Cambridge University Press, 1983.

———. *Thus Spoke Zarathustra.* In *The Portable Nietzsche,* trans. Walter Kaufmann, 103–439. New York: Penguin, 1968.

Nunes, Mark. "Sex, States, and Nomads: Comments on Julian Dibbell's Rape in Cyberspace." 28 April 1997. Available: http://www.dc.peachnet.edu/~mnunes/pres_95.html.

Perniola, Mario. *Enigmas: The Egyptian Moment in Society and Art.* Trans. Christopher Woodall. New York: Verso, 1995.

Ricoeur, Paul. *Freud and Philosophy.* Trans. Denis Savage. New Haven: Yale University Press, 1978.

Stelarc. "Stelarc Official Web Site—Australia." Available: http://www.merlin.com.au/stelarc/index.html.

Stivale, Charles J. "'Help Manners': Cyber-Democracy and Its Vicissitudes." 11 July 1996. Available: http://wwwpub.utdallas.edu/~cynthiah/lingua_archive/help_manners.html.

Stone, Allucquère Rosanne. *The War of Desire and Technology at the Close of the Mechanical Age.* Cambridge, Mass.: MIT Press, 1995.

Turkle, Sherry. *Life on the Screen: Identity in the Age of the Internet.* New York: Simon and Schuster, 1995.

———. "Who Am We?" *Wired,* 11 July 1996. Available: http://www.hotwired.com/wired/4.01/features/turkle.html.

14

Songs of Thy Selves

Persistence, Momentariness, Recurrence, and the MOO

Michael Joyce

Until recently, when its host MOO, Brown's Hypertext Hotel, went down for renovations and redesign, you could find the following in the Hi-Pitched Voices wing of that space:

```
Anne's Work Room

You see a large sun-lit room looking out over a rambling
  English garden. The windows are open and the smell of
  honeysuckle wafts in on the warm spring air. There are
  three large wooden desks. Two are covered with half-
  finished bits of code, books, papers, jottings of
  stories and poems, pictures. One is kept clear and
  here, neatly stacked, is the current work-in-
  progress. This desk has a green leather top and
  several deep wooden drawers. It magically keeps track
  of anything written on it.

Obvious exits:
out => Hi Pitched Voices
You see scribbles here.
Anne is here.
```

The poet, hypertext writer, literary feminist, activist, teacher, and computer scientist Anne Johnstone, whose room this is, and who indeed from the objective claim of the language of this space still stands here midst her scribbles and the smell of honeysuckle, and who I once hugged (tightly, in sweet regret for what might have been, bidding her goodnight after dropping her off at a Days Inn on a winter night in Poughkeepsie), died in her riverside cottage in Orono, Maine, in the company of her circle of women friends, strangers, hospice healers, and her Scots mother at a very young age, not yet forty-five, of a rapid and rapacious cancer on February 28, 1995.

I begin this writing in our garden overlooking the Hudson. There is in this sentence both a preliminary sense, as if something must happen, and an inevitable setting that most readers bring to it. The latter requires amplification: is it the Hudson in Manhattan, a place of high palisades, commerce, bustle? the Hudson of the Highlands, the green shoulders of Storm King and Breakneck opposite? or the placid Hudson below Albany, muddy and wide?

The amplification that the sentence requires is, of course, an aspect of the preliminary sense that something must happen. I am drinking a ginger beer in a brown bottle but also the sweet dregs of morning coffee in a robin's egg blue mug and tapwater from a fluted water glass. I have amplified the setting inward, making a proximate geography. Not more than thirty feet from me a concrete Tibetan Buddha sits in a weedy sprawl of jade next to a bank of wild mustard along our neighbor's wire fence.

The connection between the preliminary sense and the inevitability is a mortal space. A much overlooked aspect of death is that it is, so far as we are given to apprehend it in our conscious lives, solely preliminary, in almost no sense (except the preliminarily elegiac) the inevitable closure we attribute to it. (A sailboat crosses through the gap between two houses, one of clapboard, another eighteenth-century brick, which is my vista onto this sunny June afternoon.)

We fill this mortal space with the living of our lives, although like a hole in the sand it is never filled or rather ever filled, since we do so in recurrence, wave upon wave.

It isn't necessary that the sentence be true in either the preliminary or the inevitable sense for the space of mortality to form. Fiction, of course, both accomplishes and fills this space, more or less finally, though the reader of a print text can see it as ever filling in repeated readings or memory, while the reader of an electronic fiction must inevitably see closure as transitory.

Likewise the space between two related ideas has its own mortality, its own preliminary and inevitable states, although these are not a polarity, rather a consistent relationship, a tensional momentum in Carolyn Guyer's phrase. Take two ideas thus far opened here: the idea of the tensional momentum of the preliminary and the inevitable and the idea of language as embodied proximate geography. One of them—ironically the latter—is less fleshed out. There is a claim here, hardly made as yet, that we live our deaths by somehow symbolically mapping space (through language, image, movement, even the implied spatiality of sound that creates through interval) in relation to our bodies. Death is a condition of suspended preliminarity. One table in Anne's Work Room is kept clear and here, neatly stacked, is the current work-in-progress. This paragraph, as in waves it seems to close the gap between the two ideas that it isolated and emphasized, might seem to be squandering the space it creates and describes (especially now as I take this prose to a more analytic and self-reflexive vein, although the appearance of a self in this parenthetical weave—an "I" who takes this paragraph "there" and who drinks ginger beer on a splendid Sunday afternoon along the Hudson—in some sense opens it again).

It isn't necessary that Anne Johnstone be present in her workroom for her character to continue to inhabit the mortal space between the preliminary and the inevitable. (Though I am truly in our unkempt and lush garden and there is birdsong—squawks, cheeps, bottle songs, and recurring trills—a jet overhead on its approach to Newburgh, powerboats along the river, the occasional melancholy and tentative song of baritone chimes when a fresh breeze climbs up the sunny incline to the sunbaked house where they hang outside our mudroom window, and as well occasionally the silence of sailboats tacking on edge or white moths fluttering in the mustard flowers when I look up.) In truth, as a MOO devotee, perhaps a techie impatient with my chimes and moths and blue mug, surely knows, someone (a wizard in early instantiations of MUD and MOO cores, though lately likely to be called anything: janitor, proprietor, citizen, concrete Buddha) could inhabit the Anne Johnstone character, revive her.

But it was ever so. These things happen. A letter arrives from a dead woman a day or week beyond her death. We revive her, see her in the mixed and mortal time between the preliminary and the inevitable, create for her a proximate geography (she held her pen so, sat at her computer in the wood-paneled cottage down a slope beside the river—I imagine this, I was never there, she could have written from the Days Inn, could be there still, children believe as much, the mad, the MOO devotee). I am not disinclined to agree with our devotee, the breeze springs up and magically the full sails of a Hudson River sloop, the *Clearwater,* appears rounding the point at the Bottini Oil Company tanks in New Hamburg. The baritone chimes are the Gregorian model from Woodstock Chimes. I recall a large sunlit room looking out over a rambling English garden. My name is Anne.

"In the future she wouldn't have to die, not really. When MOO's become graphical, where there's really VR, there could be an avatar of her. You could revive her, move her through the space," says the MOO devotee. "You wouldn't need language, words get in the way."

O this I will not agree with. We have to die, this is the meaning of words. Unpleasantly or no, I believe that in the interregnum of the MOO it is language that creates the proximate, mortal space between the preliminary and inevitable. It is probably my folly, certainly my garden (though rented, we own this space in observation and action, our landlords, the Cullens of Rabbit Island on the Hudson near here, are mere wizards). I have cast the MOO Devotee out, toaded him, in MUD parlance (though why I have made the devotee a he, or why invited him here unless I needed his residence I do not know, unless, of course he is me, made in my form, cast from my garden: these are old myths). The concrete Buddha looks on serenely, his back to the river.

Words are the way, I want to say. (I rhyme like baritone chimes.) And yet I know, in the way someone watches water slip through sand, that words are being displaced by image in those places where we spend our time online; know as well that images, especially moving ones, have long had their own syntax of the preliminary and the inevitable. Time-based forms, whether music or filmic, depend on their inability to fill out the space they create and open.

Perhaps the MOO is more a film than a sestina (Anne Johnstone's form, she wrote a sequence of hypertext sestinas, what would seem an impossible form, and yet in this linked space every line seemed to double its sense of inevitability and recurrence, the frame filling with the possibility of the line, draining with the traversal of each link, all of it gathering again at the end in the eddy of envoy). But if a film, of what kind? Something like Godard's in the golden years, where the cast would arrive in the space of an idea, improvise, neither he nor they knowing where the camera went or what would follow this, what precede it. Elsewhere I have written of the MOO that "attention is always elsewhere, it is a distraction, a disposition of self, the confusion of the space for the occasion" (311). The actors join the audience in seeing the film as it results; the film as it results joins the director with the actors in seeing the film. This is the form of the sestina. (The MOO devotee, a techie, loses patience with this artsy talk and drifts off; he looks for action. He is soon elsewhere, in MOOspace where action inevitably follows the preliminarity of space.)

```
You see scribbles here.
Anne is here.
```

Against the commonly cited momentariness of MOO experience and the evanescence of the selves that form within it, there stands the rhythm of recurrence on unknown screens elsewhere; the persistence of certain "objects" that, like the consumerist flotsam of temporal existence (a brown bottle or a sailboat), mark the swell and surge of lives lived in body, space and time; and the mark of the momentary itself, meaning within meaningfulness not against meaningless. Thus, like any poetic text, the MOO aspires to moral discourse and to inscribe our mortality.

Movement is the mark of the momentary, and our mortality, alike. We move through moments and thus mark them with our bodies. Alison Sainsbury appropriates Hélène Cixous's *Three Steps on the Ladder of Writing* to call for an electronic writing that moves beyond the "hypertextual body and its ability to disrupt and interrupt" to "the necessity of embodied text . . . moving under, around, not arriving . . . the perfect unity of textual body and jouissance," which is to say one that inhabits the momentary and proximate space between the preliminary and the inevitable. The image Cixous has in mind in her text is the daughter of the Grimm fairy tale who each night pushes her bed aside, lets herself down through a trapdoor, goes off into the night, and wears her shoes out dancing. "In order to go to the School of Dreams," Cixous writes:

> something must be displaced, starting with the bed. One has to get going. This is what writing is, starting off. It has to do with activity and passivity. This does not mean one will get there. Writing is not arriving; most of the time it's not arriving. One must go on foot, with the body. One has to go away, leave the self. How far must one not arrive in order to write, how far must one wander and wear out and have pleasure? One must walk as far as the night. One's own night. Walking through the self toward the dark. (65)

The character on the MOO is always starting off, must always "begin to begin again," in William Carlos Williams's phrase. Late in a MOO session someone joins, and, despite whatever coherence the evening and the space has offered us, we are all once again starting off, not arriving where we began to think we were. Not just the character (our own or this late-arriving one, perhaps it is the MOO devotee coming back; perhaps it is the avatar of the concrete Buddha as Anne Johnstone) offers us a lasting presence of particularities as a strategy against the fragmentary plenitude and multiplicity that faces and effaces us. We are all just starting off, all recently arrived. Someone moves from another room to this one—one of those characters who arrives in a comet of textual introduction, dragging a tinker's bag of objects and attributes behind him as the tail (or the tale: it *is* the MOO devotee!)—and his arrival disrupts the surface of what we have settled on among us (as the comet light stirs a dark pond). His presence once again marks the momentary space between the preliminary and the inevitable. He is what catastrophe theorists call a perturbation. (Oh here let us free him from his gender and his character; we are in a MOO, after all; he is not the MOO devotee, she is the shapeshifter.)

What can happen now that she is here (not yet arrived at the dance and yet here)? We can summon hierarchical notions of meaning and power: toad her, burn her at the stake, ignore her (Judy Malloy in the early stages of Brown Kitchen would set up space inside a room at LambdaMOO and begin to tell her stories, ignoring the yelps and noise and already flowing meanings until the story made its own space, like the eddy of a contrary-flowing river). Or we can (continue to) make our own meaning within and against and in complicity with the arriving comet. Yet in the shifting current the eddy (the mortal space between the preliminary and the inevitable) has both enfolded and unfolded. What we make will never be the space of what went before; what went before was never the space we thought it was as it unwound, scrolling upward like smoke, flowing away like a river or towns on a Triptik.

The arrival of the avatar (shapeshifter, Judy Malloy, Anne Johnstone, our friend the MOO devotee) offers an occasion of coherence. Coherence in this sense is very close to what the catastrophe theorists (characters who arrived not long ago in a comet's spew) mean by a singularity or phase shift, that is, a recognizable change in which something amorphous takes on form defined by its own resistance to becoming anything other than its own new form. Coherence is making sense for oneself and yet among others.

A room in a MOO is not the meaning of what it contains but the space of its coherences. You choose to understand the flow of talk and action that you are within, choosing also from what perspective (what self) to view it. Following Cixous, Alison Sainsbury suggests that we must go beyond the self in order to reclaim the active self as something newly whole. Coherence can be seen as partially meaningful patterns emerging across a surface of multiply-potentiated meanings into something newly whole: the eddy in the river, a brown bottle, a desk with a green leather top that "magically keeps track of anything written on it." Coherence distinguishes itself not against but within other possible coherences, in the recurrent flickering of meaningfulness not meaninglessness.

```
You see scribbles here.
Anne is here.
```

Days have passed; it is (Tuesday) too steamy to sit by the river, although the Buddha has not moved. I want to walk out after dark, a walk through the self toward the dark.

Let us talk of movement for awhile and see where that gets us. I have often made the claim that the MOO, as a variety of hypertext, is essentially spatial. Yet how to reconcile this with the contrary and obvious claim that the MOO is temporal? What we are looking for here is the mark of the momentary, the space between the preliminary and inevitable. Recurrence, I have said elsewhere, is the memory of spatial form. How something can mean, not what it does mean. The repetition of a phrase, "Anne is here" for instance, marks both an indeterminate space (unless you happen idly or paranoically to have decided to count the lines or paragraphs or pages after a certain phrase in case it should appear again) and an indeterminate time (you might likewise have either used a stopwatch, or looked for her, or decided this phrase is a narrative: first she was there, then she was reported dead, now he says she is here but means something else by it; this succession of temporal measures—metric, experienced, narrative—is the stuff of the MOO experience). The MOO repeats itself both determinately and indeterminately. The rooms are (largely) determinate, characters less so. The formulaic aspect of many actions (so-and-so says such-and-such) is determinate, the substantive (Devotee says, "This is philosophical claptrap") is not.

Yet these distinctions do not sustain themselves. Moving from room to room in some MOOs (Hi-Pitched Voices for instance) we accrue meaning both directly (reading the walls as it were) and indirectly (from the customary interchanges among characters). Rooms thus oscillate between determinate and indeterminate states according to the order in which we encounter them and in whose presence (or absence: Anne is here) we do so. Likewise actions, which seem most indeterminate (Devotee snarls, scorns, turns into Shapeshifter; Catastrophe Theorist rains and rants and reigns) become by their recurrence markers of a constant set of changes that cohere as the character who enacts them.

Even the measures of temporality oscillate, as in Einstein's famous trope of the relativity of time for someone seated on a hot stove versus someone seated on a lover's lap where experienced and metric time interact; or again how in their metrics some video games observe narrative time (or what we could call Disneyland time, at three-quarter scale or less), compressing thirty-minute halves of a soccer game into twelve. The coherences of spaces, characters, and temporality, while simultaneous, are not always contemporaneous or coterminous in the MOO. Not just the MOO experience but its coherences (what we might call "the meaning") are distributed, duplicated, alternately experienced across the screens of its participants as if an audience watched a film projected kaleidoscopically.

Perhaps the MOO is more a film than a sestina, I suggested. Interactive filmmaker Grahame Weinbren considers the question of a similar system of indeterminate narratives and explorations. In the "interactive cinema," he says, "All and any loose narrative ends will never be knotted. . . . If a viewer navigates through a mass of material, some of it will be seen and some won't, and surely some of what isn't seen earlier will raise issues that remain unresolved in what is seen later."

How what is not seen might raise issues that we later recognize as unresolved rehearses the connection already made here between the preliminary and the inevitable within a mortal space. We all die of what at some point were unknown causes and in so doing resolve what we did not see before. We attribute meanings to the cessation of experience; this is the nature of closure. As a slogan we might say that moral resolutions emerge from mortal recurrences. That is, we attribute morality to the mortal space between the preliminary and the inevitable.

There is a mechanism for this, although we are unlikely to attribute to it the cultural certainty with which certain narrative theorists attribute a moral frame to the beginning-middle-end coherence. "A system can be sensitized to repetition," says Weinbren, "either so as to avoid it, or so that as soon as repetition starts the viewer is offered the opportunity to enter a structurally different region, a territory of culmination or summary." People are not systems (systems are what people are not), and yet we do program ourselves to avoid repetitions (it is Wednesday now, back in the garden, and across the road there is the sound of a gas-powered trimmer that I ignore, nearer a blue jay's hoarse, repeated, jug-deep song); do consider that some recurrent experiences (the seven danger signals of cancer for instance, or love) signal a territory of culmination, that is, ever after perhaps happily.

However, while it is true as Weinbren says that "in general terms, a map of territory covered can be kept by the system, and once a certain area has been explored, closure possibilities can be introduced," this seems where, despite our most fervent, earnest protest (and, Horatio, despite all your cognitive sciences), people and systems differ. We don't "get it." Don't "work our way through" the territories of the heart until we reach a place where suddenly we can "see clearly," a clearing where we know "what's what, what's up, what matters."

We do, of course (do I contradict myself? well then I contradict myself), but at the level of longing, of desire. We fill the mortal space between the preliminary and the inevitable with a moral landscape of our own (re)making, a succession of coherences (as it was in the body, is in the MOO, and ever shall be in memory). The moral landscape remains a Hesiodic space, the world (even the MOO, the blue jay, or the concrete Buddha) is held by the glue of desire. "All this is to say," says Weinbren, emerging to a clearing where he can see what's what, what's up, what matters, "that despite its need for an opened narrative, closure cannot be banished from the Interactive Cinema. Remove the imminence of closure and we begin to drain cinema of desire. Closure must be recast in a more radical light."

So we shall. Let us consider the most radical light, the light of lights, and look on the moral landscape with a theologian's eyes. There (or at least in this one theologian, Gabriel Marcel), the truth is that the dead live, Anne is here, you see scribbles:

> The spirit of truth bears another name which is even more revealing; it is also the spirit of fidelity, and I am more and more convinced that what this spirit demands of us is an explicit refusal, a definite negation of death. The death here in question is neither death in general, which is only a fiction, nor my own death insofar as it is mine . . . ; it is the death of those we love. They alone are within our spiritual sight, it is they only who it is given us to apprehend and to long for as beings, even if our religion, in the widest sense of the word, not only allows us but even encourages and enjoins us to extrapolate and proclaim that light is everywhere, that love is everywhere, that Being is everywhere. (147)

It is a commonplace to think that MOO pranking, MOO terrorism, MOO marauding and worse (see Dibbell) harken back to the days (they continue of course) of the MUD and its roots in adventure games and especially in D&D, where boys killed boys in the guise of knights and dragons and orcs and theologians. That this commonplace happens to be true does not keep us from recognizing that the MOO is a place of death's denial, both in the commonplace slaughter of imaginary (though no less dangerous) boys and the riverside workroom of the poet as well. Marcel's notion of denial as fidelity, while couched in negation, is imbued with a positive charge of desire. According to Marcel our task (our life) is "to apprehend and to long for" the dead "as beings."

It isn't immediately obvious, even on the MOO, how to engage in a language of denial and desire. You can't simply say the dead live (Anne is here), nor is Marcel's theology an abracadabra one in which we do so.

Fidelity requires repetition and recurrence; it is in this sense only "true to life." True *to* life is true *through* life (this is another, equally sloganeering, version of the earlier slogan "moral resolutions emerge from mortal recurrence"). As Cixous suggests (I have long thought that at some early age she read Marcel, based on little more than intuition and certain echoes in their language), "We live bizarrely clinging to the level of our age, often with a vast repression of what has preceded us: we almost always take ourselves for the person we are at the moment we are at in our lives" (66). This is a misappropriation, a reversal of the relation between the preliminary and the inevitable, seeing oneself as arrived rather than not arriving. Cixous, longing as always, seeks to reverse the reversal: "What we don't know how to do is to think—it's exactly the same as for death—about what is in store for us. We don't know how to think about age; we are afraid of it and we repress it" (66). What is in store for us is another way to describe the preliminary; we think about age by marking the momentary, through recurrence. "Writing," says Cixous, "has as its horizon this possibility, prompting us to explore all ages" (66).

Here a self-satisfied MOO devotee could stop, note (even exult) that MOOspace is little else than a horizon of possibilities, an exploration of seven times seventy ages of woman and man from Jesus to Jacques to Janis (or Scott) Joplin, from Mohammed to Melanchtha to Mrs. Ramsey. Yet we likewise need to stop here and consider what seems the evident drabness and regular debasement (as in the smartass sequence from Jesus to Joplin) of MOO language. The debasement of language in electronic texts leads some cultural commentators and critics to argue that theory such as this essay overstates the ordinariness or worse of the language of MOOspace, the Web, and so on. In fact, canonical critics might argue that the argument here is carrion criticism, parasitically feeding on waste in order to puff up a claim of transcendence and the poetic. The MOO (all of electronic text) is faulted for not having a language worthy of eternity. This is of course a macho claim; electronic space doesn't have balls—not brass but *aere perennius* (more lasting than bronze, in Horace's phrase)—enough to stand up to history. Canonical criticism is a claim made by imaginary boys not different except in their pedigree from MOO marauders.

Such challenges are not so much overthrown (it is the same game, a joust of imaginary knights: these canonical critics are strawmen, in my own image and of my own making as much as the MOO devotee) as much as seen through, both in the sense of persistence and the transparent sense of boys in costumes. It isn't necessary to see the MOO, as Erik Davis does, as the "apotheosis of writing . . . nonlinear texts in many ways more marvelous than the precious literary experiments" (64). Instead we can see

MOO writing as something much more routine and commonplace (what is more commonplace than imaginary nights? more commonplace than desire and denial, than religion?). Cixous in the "School of Dreams" section of *Three Steps on a Ladder of Writing* sees the commonplace "face of God, *Which is none other than my own face,* but seen naked, the face of my soul" (63). Cixous's face of God is a stack of tender lies, what in speaking of electronic text my student DeAna Hare calls "kitchen talk." The recurrent mark of the momentary (ungendered, though most common among women) becomes, however briefly, a coherence for Cixous:

> The face of "God" is the unveiling, the staggering vision of the construction we are, the tiny and great lies, the small nontruths we must have incessantly woven to be able to prepare our brothers' dinner and cook for our children. An unveiling that only happens by surprise, by accident, and with a brutality that shatters: under the blow of the truth, the eggshell we are breaks. Right in the middle of life's path: the apocalypse; we lose a life. (63)

It is exactly in the commonality of MOOspace, the noise and actions, the kaleidoscopic projections, the constant replacement that I have elsewhere suggested (227) is the characteristic of electronic text, which, rather than in the text itself (the content, to use that reprehensible word with which technocommerce would end art) that we lose and gain our lives. MOOtalk is kitchen talk, a gathering of aunts in which, as Grahame Weinbren characterizes the intermingling narratives of interactive cinema, "several story lines continue until one, some, or all of them end." Weinbren too begins with water (Rushdie's *Haroun and the Sea of Stories* and Barth's *Tidewater Tales*), bringing him to a place where "numerous Diegetic Times are constantly flowing forward, many narratives operating in time simultaneously whether or not the viewer encounters any particular one . . . potential narrative streams, elements themselves unformed or chaotic, but taking form as they intersect, gaining meaning in relation to one another."

Let us end then in water as well and (for awhile, until the end's end) a series of dead white men, flickering by as if on a screen of light. I suppose one could begin to end with Eliot (I have regularly been accused of being a modernist) whose river in the *Four Quartets* is "a strong brown god . . . ever, however, implacable / Keeping his seasons and rages, destroyer, reminder / of what men choose to forget," and who "tosses up our losses" (35–36). But instead, longing as always, let us heed the call of desire and denial (do you wonder where this hortatory "us" has entered? or did you, determinate, mark the paragraphs since I turned to this diction, creating you in my own image? It is a proximate geography: under a lapsing morning haze the river hesitates ripe and fragrant between tides, not flowing either way, as a breeze strikes up out of the humid overcast and the rhythm of the birdsongs hastens, the jay turned now from jug-song back to the more ordinary grating call).

Against the commonly cited momentariness of MOO experience and the evanescence of the selves that form within it, there stands the rhythm of recurrence on unknown screens elsewhere; the persistence of certain "objects" that, like the consumerist flotsam of temporal existence (a brown bottle or a sailboat), mark the swell and surge of lives lived in body, space and time. The mark is the mark of the momentary itself, meaning within meaningfulness not against meaningless. The flicker on so many screens elsewhere is denial and desire together, the oscillation of the preliminary and the inevitable, a serial recognition (temporal, recurrent, harmonic because at intervals) of our shared mortality and persistence. In "Crossing Brooklyn Ferry," Walt Whitman wrote:

> It avails not, time nor place—distance avails not,
> I am with you, you men and women of a generation, or ever so many generations hence
> Just as you feel when you look on the river and sky, so I felt,
> Just as any of you is one of a living crowd, I was one of a crowd,
> Just as you are refresh'd by the gladness of the river and the bright flow, I was refresh'd
> Just as you stand and lean on the rail, yet hurry with the swift current, I stood yet was hurried. (159)

Writing, Whitman puts us at Cixous's horizon of possibility, "prompting us to explore all ages." This is a conscious attempt at a proximate geography, a claim for the transcendence of the virtuality of language over the mortality of the body. The claim for transcendence lies as much in the recurrent actions of the "you" (it is us) who are addressed and summoned to the spirit of fidelity as it does in the language itself. We persist in the mark of the momentary, the endless starting off of denial and desire alike, the commonplace and anonymous rhythm that Samuel Beckett summons in "From an Abandoned Work":

> Oh I know I too shall cease and be as when I was not yet, only all over instead of in store, that makes me happy, often now my murmur falters and dies and I weep for happiness as I go along and for love of this old earth that has carried me so long and whose uncomplainingness will soon be mine. Just under the surface I shall be, all together at

first, then separate and drift, through all the earth and perhaps in the end through a cliff into the sea, something of me. (160)

Anne is here. Let us end then. I write this in preliminary apprehending and inevitable longing for Anne Johnstone, who lives, no less in these words than in her embrace impressed once upon the body of my memory.

Works Cited

Beckett, Samuel. *The Complete Short Prose: 1929–1989.* Ed. S. E. Gontarski, New York: Grove Press, 1995.

Cixous, Hélène. *Three Steps on the Ladder of Writing.* Trans. Sarah Cornell and Susan Summers. New York: Columbia University Press, 1993.

Davis, Erik. "It's a MUD, MUD, MUD, MUD World." *Village Voice,* February 22, 1994, 42–43.

Dibbell, Julian. "A Rape in Cyberspace; or, How an Evil Clown, a Haitian Trickster Spirit, Two Wizards, and a Cast of Dozens Turned a Database into a Society." *Village Voice,* December 21, 1993, 36–42.

Eliot, T. S. *Four Quartets.* New York: Harcourt, Brace and World, 1943.

Guyer, Carolyn. "Along the Estuary." In *The Millennial Muse,* ed. Sven Birkerts. St. Paul: Graywolf Press, 1996.

Joyce, Michael. "Moo or Mistakenness." *Works and Days* 13, nos. 1–2 (spring–fall 1995): 305–17.

Marcel, Gabriel. "Value and Immortality." In *Homo Viator.* Trans. Emma Craufurd. New York: Harper and Brothers, 1963.

Sainsbury, Alison. Email to the author, October 11, 1994.

Weinbren, Grahame. "In the Ocean of Streams of Story." *Millenium Film Journal* 28 (spring 1995): Interactivities, 15–30. Available at: http://www.sva.edu/MFJ/GWOCEAN.HTML.

Whitman, Walt. *Leaves of Grass.* Ed. Harold W. Blodgett and Sculley Bradley. New York: New York University Press, 1965.

Appendix

MOO Central

Educational, Professional, and Experimental MOOs on the Internet

Jeffrey R. Galin

Introduction

In this appendix I have identified as many active educational MOOs as I have been able to find, grouping them into categories. By active, I mean MOOs that are used regularly during the school year and have a substantial base of growing projects under way. I define educational MOOs as those whose primary function is to serve groups of students and teachers for course-related work. While the majority of the text-based virtual realities listed here are MOOs, there are also a few MUDs, MUSHes, and MUSEs. The fundamental differences among these classes of virtual worlds are the command sets they use. If you are unsure of a command on any of these systems, all provide help or tutorial facilities that are clearly identified upon entering. Make note of these resources and you should seldom have problems. Because some social and professional MOOs welcome teachers and their students and provide researchers opportunities to hold professional events, experiment with new interfaces, and develop new objects for educational purposes, I have included categories for professional, experimental and programming, and a few social MOOs.

The full list of categories is

General educational MOOs

University specific MOOs

OWLs (Online writing centers)

Foreign-language MOOs

K–12 MOOs

Professional MOOs

Experimental and programming MOOs

Social MOOs

Selected educational web sites on MOOs and MUDs

MUD/MOO/MUSH Libraries

All links in this appendix were working as of August 1997.

General Educational MOOs

ATHEMOO is a professional community that invites theater aficionados to exchange information, attend meetings, and learn about MOO technology and its uses for classroom and other projects. The Association for Theatre in Higher Education (ATHE) has developed this MOO with the support of the University of Hawaii. ATHEMOO is a GNA (Globewide Network Academy) MOO.
Telnet: moo.hawaii.edu 9999

AussieMOO is jointly sponsored by Charles Sturt University and Paideia, a university on the Internet since 1992. It was developed as an open-styled, experimental, and research MOO for social interaction, conferencing, computer-supported cooperative work (CSCW), lifelong education, object-oriented programming, experimental psychology and philosophy.
WWW: http://silo.riv.csu.edu.au/AussieMOO.html
Telnet: farrer.riv.csu.edu.au 7777

CollegeTown was created to provide a virtual workspace in the sciences and humanities outside of the office and classroom. This MOO fosters cross-disciplinary collaboration and welcomes academics from all levels and disciplines to apply for residence. Collegetown is structured to support small seminars, workshops, and collaborative research groups for all ages. The only requirement for membership "is a serious commitment to academic projects." CollegeTown is a GNA (Globewide Network Academy) MOO.
WWW: http://www.bvu.edu/ctown
Telnet: galaxy.bvu.edu 7777

Connections is a virtual learning environment. Classes of all kinds and all levels can make arrangements to use this space; so can those wishing to set up or participate in other kinds of learning projects. Explore this space as a guest, and if you find that you're interested in using Connections, write to Tari: tari@ucet.ufl.edu.
Telnet: connections.sensemedia.net 3333

DaedalusMOO is less structured than several other Moos. It allows teachers to bring their students to collaborate with others and discuss use of

Daedalus Integrated Writing Environment (DIWE) (http://daedalus.com/index.html) software. Daedalus also hosts several special projects, such as the Cyberspace Writing Center Consultation Project (Writing Works), a virtual-reality writing center.
WWW: http://fur.rscc.cc.tn.us/Cyberproject.html
Telnet: logos.daedalus.com 7777

DaMOO is an educational webbed MOO, supported by the Learning Resource Center at California State University, Northridge. DaMOO is host to a range of academic, professional, and social meetings. In April 1996 the Coconut Cafe was christened to serve as a meeting place for educators at community colleges as part of the TCC-L On-Line Conference, "Innovative Instructional Practices." As a relatively young MOO, the administrators invite you to help them explore and develop this environment.
WWW: http://lrc.csun.edu 8888/
Telnet: lrc.csun.edu 7777

Diversity University (DU) is designed as a virtual campus for teachers to bring their classes of students to. This environment includes many educational models from traditional to constructivist and allows for both personalization and specialization of an individual teacher's facilities while still offering the technical and organizational expertise of the volunteer staff. DU is also a nonprofit organization that provides educational services on DU and other MOOs, such as the Online Educators' Resource Group (OERG). DU is a GNA (Globewide Network Academy) MOO.
WWW: http://www.du.org
Telnet: moo.du.org 8888

Education Multimedia Moo at Monash University (EdMOO) is primarily for teacher collaboration and research. Virtual meetings are scheduled from time to time; however, participants may drop in at any time to chat or create or play with virtual objects. EdMOO has been enhanced with various features using the MOO environment and the specifically designed web-based drover client software. Drover provides the ability to "show" or "whisper" WWW pages to others in a room, to display files or graphics from Internet sites to all players or to one player, to play sounds and movies from Internet sites (as supported by Netscape). It has a full mouse/icon/button driven interface.
WWW: http://edx1.educ.monash.edu.au/projects/moo/index.html
Telnet: edx2.educ.monash.edu.au 7777

GNA-Lab serves as the central MOO of the Globewide Network Academy, which describes itself as "an educational and research organization dedi-

cated to providing a competitive marketplace online for distance learning courses and programs." The mission of the GNA Net is to create a comprehensive source of information in the form of a "central listing of online courses and degree programs worldwide."
WWW: http://uu-gna.mit.edu:8001/uu-gna/index.html
Telnet: gnalab.uva.nl 7777

Grassroots MOO is designed for students and educators to be able to participate in a "virtual neighborhood," in which they can explore, interact, and learn. This public MOO welcomes registered characters of students and teachers to help construct rooms for hands-on experience and to develop web pages to showcase their work on Grassroots MOO. Make sure to check out the list of educators' links, including a listing of affiliated web sites.
WWW: http://rdz.stjohns.edu/grassroots/
Telnet: rdz.stjohns.edu 8888

IPL-MOO (Internet Public Library MOO) is being developed as a public library for the Internet community, and is a live, real-time extension of the Internet Public Library. This MOO is intended as a model and laboratory for librarians and MOO folk to explore the possibilities of library services in MOO and virtual-reality environments. IPL-MOO is a GNA (Globewide Network Academy) MOO.
WWW: http://ipl.sils.umich.edu/
Telnet: ipl.sils.umich.edu 8888

Lingua MOO was created to serve primarily the University of Texas at Dallas Rhetoric and Writing program and the School of Arts and Humanities. It serves as both a learning environment for their students and a broader community for research and collaboration on projects situated at the intersection of Arts and Humanities and electronic media. Lingua MOO is also home to an international network of researchers in these areas and supports links with other educational MOOs in the GNA Net (Globewide Network Academy).
WWW: http://lingua.utdallas.edu/
Telnet: lingua.utdallas.edu 8888

Meridian is a global community shared by people interested in virtual travel, cultural interaction, friendship, and learning. Here, people from around the world represent real places they know intimately, for others to experience and enjoy in this medium. Meridian is a GNA MOO.
Telnet: sky.bellcore.com 7777

VOU (Virtual Online University, Inc.) is a nonprofit corporation that engages in two educational missions: Athena University for higher education (http://www.athena.edu/Athena-Home.html) and Athena Preparatory Academy (http://www.athena.edu/apa.html) for preuniversity education as well as consultations for corporate trainers and educational institutions seeking an effective distance educational-delivery system. VOU represents a new model for education and training by providing real-time, computer-mediated classes from an electronic campus. Anyone interested in examining the MOO environment for distance learning must take a look at VOU. They have a wealth of information available for downloading.
WWW: http://www.athena.edu/
Telnet: athena.edu 8888

University-Specific MOOs

AcademICK is actually a TinyMUSH located at the Computer Writing and Research Labs (CWRL) at the University of Texas at Austin. This research TinyMUSH was set up to explore its educational possibilities. AcademICK's topology is based on the hub, or nexus, model: characters connect to "The Forum," the central space from which a series of arches lead to different areas of the project.
WWW: http://www.cwrl.utexas.edu/researchpoints/ick.html
Telnet: writing.en.utexas.edu 7777

IATH-MOO is sponsored by the Institute for Advanced Technology in the Humanities, at the University of Virginia. While this MOO is host to a range of professional conferences and class usage, it also features a multiple-narrative fiction and a text-based game called M'Cluhan's Pines. As a fiction, it is the story of the abandoned home of a dead woman, the ambiguous "crazy lady" of the neighborhood. The story is embedded in the rooms, objects and details of the house, and the game aspect is to find out the woman's inner identity and domestic history.
WWW: http://jefferson.village.virginia.edu/~482s95/mms7s/pines.html
Telnet: hero.village.virginia.edu 8888

MiamiMOO is a collaborative project at Miami University of Ohio involving the Departments of Classics, Religion, Educational Psychology, Art, and Architecture in collaboration with visitors from other institutions. As an academic MOO, its purposes are to provide learning experiences that cannot be achieved in traditional classrooms, to host models of important historical and religious buildings that no longer exist, and to document the interactions that took place in those places.

WWW: http://miamimoo.mcs.muohio.edu/
Telnet: moo.cas.muohio.edu 7777

MOOville is dedicated to teaching and research at the University of Florida. As an essential component of the UF/IBM Writing Project's Networked Writing Environment (NWE), MOOville entertains up to three thousand students in over seventy classes a semester. Teachers use MOOville as a teaching space for a wide range of classes, including Writing about Literature, Writing with Media, Fiction and Poetry Writing, and Argumentative and Expository Writing.
Telnet: moo.ucet.ufl.edu 7777

El MOOndo has been a MOO for K–12 types, the graduate students of Linda Polin in the Graduate School of Education and Psychology at Pepperdine University, and "people we like and let in." The theme is pastoral, wandering through forest paths, underneath waterfalls and through a country village. A few parts are in Spanish and English. Dr. Polin says that you "may apply for a character if you think El MOOndo feels like a good place to live."
WWW: http://moon.pepperdine.edu/~lpolin/MOOStuff.html
Telnet: gsep.pepperdine.edu 7777

PennMOO is a virtual learning center for the University of Pennsylvania, which is sponsored by the Educational Technology Services, School of Arts and Sciences Computing. Make sure to check out "What's Happening at PennMOO" on their web site.
WWW: http://www.english.upenn.edu/PennMOO/home.html
Telnet: moo.sas.upenn.edu 7777

PSCS VEE (Virtual Educational Environment) is Puget Sound's virtual extension of their Community School, founded in 1994. The founding principle of the school is that learning is best fostered by self-motivation, self-regulation, and self-reflection. The curriculum is determined by the interests of the students, with equal status given to all pursuits. PSCS believes "the best education insures the freedom of students of all ages to think for themselves and to control their own destinies, structuring their time and activities as they best see fit."
WWW: http://www.pscs.org/
Telnet: moo.speakeasy.org 7777

TecfaMOO identifies itself as a "Virtual space for educational technology, education, research and life at TECFA, School of Psychology and Educa-

tion, University of Geneva, Switzerland." TECFA hosts one of the most comprehensive educational MOO undertakings available to date on the Internet. Like many other university MOOs, it is rather specialized: it serves primarily students in the postgraduate "STAF" diploma in the educational-technology program. TecfaMOO is a GNA MOO.
WWW: http://tecfa.unige.ch/tecfamoo.html
Telnet: tecfamoo.unige.ch 7777

VUW is the virtual campus at the University of Waterloo. By connecting to VUW, students can interact with each other and hold discussions specific to English 210e. Students may also choose to use the MOO as a means of getting to know each other and pass less stringent leisure time.
WWW: http://watarts.uwaterloo.ca/~camoock/moo1.html
Telnet: watarts.uwaterloo.ca 7777

Walden Pond MOO welcomes guest logins but reserves the majority of registered characters for University of Alberta and University of Hawaii students. Others who are interested in their educational mission will be favorably considered.
WWW: http://www.ualberta.ca/CNS/PUBS/WaldenPond.html
Telnet: olympus.lang.arts.ualberta.ca 8888

OWLs (Online Writing Centers)

VWCMOO (Virtual Writing Center) is a component of Salt Lake Community College's impressive writing-center WWW site. It exists to allow students, instructors, and the community access to information about writing, tutoring, composition teaching, and computer-assisted writing instruction. This virtual writing center works in league with the "real-life" writing center.
WWW: http://www.slcc.edu/wc/welcome.htm
Telnet: bessie.englab.SLCC.edu 7777

ZooMOO of the On-line Writery at the University of Missouri is an online environment for writers. While they have to give first priority to University of Missouri students, they welcome the chance to talk with anyone from anywhere. There is a host of communications and information resources available on their web site, and they offer HTML tutoring.
WWW: http://www.missouri.edu/~wleric/writery.html
Telnet: moo.missouri.edu 8888

Foreign-Language MOOs

LittleItaly MOO has a social orientation that is sponsored by the dissertation lab of the Computer Science Department at the Universita' Statale di Milano. Users are encouraged to explore and chat in Italian, but most users speak some English. When connecting, keep in mind that Italy is five hours ahead of U.S. time. The user population is about 70 percent Italian and 30 percent non-Italian.
WWW: http://little.usr.dsi.unimi.it:4444
Telnet: little.usr.dsi.unimi.it 4444

MOOsaico is built around the theme of cultural connectivity. Though not strictly an educational MOO where teachers might bring their students, MOOsaico encourages users to build multilingual facilities (particularly in Portuguese) and to develop "functional objects" that encourage participation, liberal connectivity among MOO rooms, and a proliferation of public spaces inviting to all users.
WWW: http://mes01.di.uminho.pt/RVirtual/AMB_VIRT/amb_virt.en.html
Telnet: moo.di.uminho.pt 7777

MundoHispano MOO is being developed by Lonnie Turbee (aka Colega) with the support of the Department of Languages, Literatures and Linguistics and ERIC's AskERIC project, both at Syracuse University. MundoHispano is a community of native speakers of Spanish from around the world, teachers and learners of Spanish, and computer programmers, all of whom volunteer their time and talent to make this a dynamic virtual world.
WWW: http://web.syr.edu/~lmturbee/mundo.html
Telnet: moo.syr.edu 8888 or europa.syr.edu 8888

schMOOze University is a small, friendly college MOO known for its hospitality and the diversity of the student population. It was established in July 1994 as a place where people studying English as a second or foreign language could practice English while sharing ideas and experiences with other learners and practicers of English. Students have opportunities for one-on-one and group conversations as well as access to language games, an online dictionary, USENET feed, and gopher access. SchMOOze also welcomes all people interested in cross-cultural communication.
WWW: http://schmooze.hunter.cuny.edu:8888/
Telnet: schmooze.hunter.cuny.edu 8888

K-12 MOOs

Explorenet is an experimental MOO that is part of the Virtual Academy Project at the University of Central Florida and the Coalition of Essential Schools at Brown University. Its goals are to provide a simple-enough interface for sixth graders to use within a minute of encountering it, to utilize standard PCs and Macintosh computers and to develop a networked cooperative work environment for design and development of ExploreNet worlds.
WWW: http://longwood.cs.ucf.edu:80/ExploreNet/public_html/

MicroMuse is a multiuser simulation environment based at MIT's Artificial Intelligence Lab. The system features explorations, adventures, and puzzles with an engaging mix of social, cultural, recreational, and educational content. A few examples of what is available in MicroMuse are explained by Barry Kort: "[T]he MicroMuse Science Center offers an Exploratorium and Mathematica Exhibit complete with interactive exhibits drawn from experience with Science Museums around the country. A highlight of the Mathematica Exhibit is 'Professor Griffin's Logic Quest', based on Raymond Smullyan's classical puzzles about knights and knaves."
Telnet://musenet.bbn.com

Pueblo is a text-based virtual learning community. Its developers explain that it began as MariMUSE (http://pcacad.pc.maricopa.edu/MariMUSE/MariMUSE.html) in spring, 1993, "as a partnership program between Longview Elementary School and Phoenix College. This partnership reflects Phoenix College's commitment to its community and the development of the talent of youth in its service area. The partnership is concerned with educational restructuring efforts that contribute to success of students in inner-city Phoenix."
WWW: http://pc2.pc.maricopa.edu/

SchoolnetMOO explains that its purpose "is to provide an educational and interactive environment for elementary and secondary school students." They are currently working to develop French content on the MOO.
WWW: http://www.schoolnet.ca/moo/
Telnet: moo.schoolnet.ca 7777

V-levelMOO, sponsored by Front Range Community College, was made for online classes to receive lectures in an interactive environment.
WWW: http://www.com.frcc.cccoes.edu/support/support.htm

Professional MOOs

BioMOO is a professional community of biology researchers who meet to brainstorm, hold colloquia and conferences, and explore the serious side of this new medium. All users of BioMOO are identified and can be located for further discussion via the 'whois' command. BioMOO is a GNA MOO.
WWW: http://bioinformatics.weizmann.ac.il/BioMOO/
Telnet: bioinfo.weizmann.ac.il 8888

EnviroMOO (formally prismMOO) promotes real-time discussion and networking between environmental health and safety professionals, faculty, staff, and students. EnviroMOO is operated by the University of Toledo Safety and Health Office.
Telnet: avatar.phys-plant.utoledo.edu 5555

MediaMOO, sponsored by MIT, is one of the first, and still one of the most useful, professional MOOs designed for research rather than classroom use. There is a large group of rhetoric and composition folks who meet weekly for Tuesday Cafe to discuss a wide range of issues, particularly integrating technology and writing. MediaMOO is a GNA MOO.
Telnet: purple-crayon.media.mit.edu 8888

Experimental and Programming MOOs

E_MOO is dedicated to research and advancements in the MUD and MOO community. Though open character creations are disabled, characters are available upon request. E_MOO has served all kinds of MOOers: hardcore programmers, thinkers, talkers, idealists, newbies, and of course, the ever-present idlers.
WWW: http://tecfasun1.unige.ch:4243/
Telnet: tecfa.unige.ch 4242

JHM (formerly JaysHouse MOO) describes itself as "a place where a group of people does research and development into information structures, tools for collaboration, and cohesive simulated environments." As one of the first MOOs online, JHM remains a hotbed for experimental design and development. In addition to being instrumental in developing some of the first web-based MOO coding, programmers at JHM are currently working on projects like JHM VR, modeling narrative reality, and studies in collaboration and usability. JHM, however, does not host classes.
WWW: http://jh.ccs.neu.edu:7043/

mirrorMOO is loosely based on the theme of *Alice in Wonderland,* but its purpose is to experiment with existing and future MUD technologies. An

effort is specifically being made to combine the positive contributions of both the MOO and LPmud communities.
WWW: http://www.mars.org/home/rob/mirrormoo.html

MOO2000 is an educational/programming network, sponsored by CNS Internet Services of Grand Rapids, Michigan (http://www.cns.net). MOO2000 boasts of having cyberspace's friendliest wizard staff. And they now feature "FREE CONTINENTAL BREAKFAST for all new characters!!"
WWW: http://moo.cns.net:3434/
Telnet: moo.microwave.com 2000

MOOtiny is a WOO (webbed MOO) sponsored by the University of Nottingham, Artificial Intelligence Laboratory. It is dedicated to developing web access and friendly interaction within an island theme. It uses an expansion of the web system implemented at ChibaMOO (aka The Sprawl: http://sensemedia.net/).
WWW: http://spsyc.nott.ac.uk:8888/
Telnet: spsyc.nott.ac.uk 8888

PMC-2 MOO is an outgrowth of the online magazine *Postmodern Culture.* It is sponsored by the Institute for Advanced Technology in the Humanities. PMC-2 is a virtual space designed to promote the exploration of postmodern theory and practice.
Telnet: hero.village.virginia.edu 7777

WaxWeb, a partner of the Sprawl, is a "network-delivered hypermedia project, based on David Blair's electronic film WAX or the discovery of television among the bees." WaxWeb combines one of the largest hypermedia narrative databases on the currently free Internet with a unique authoring interface that allows Netscape or MOO users to make immediate, publicly visible hypermedia links to the main document.
WWW: http://bug.village.virginia.edu/

U-MOO (University of MOO) is a virtual world designed for the interaction of different people with different views in a manner that promotes educational discussion. U-MOO is run by the School of Computer Science at the University of Windsor.
Telnet: moo.cs.uwindsor.ca 7777

Social MOOs

BayMOO is a humanities MOO whose purposes include building a virtual community based on respect, constructing interesting virtual objects, host-

ing conferences and classes, and promoting the exchange of ideas on a wide range of social, artistic, and technical subjects. BayMOO is a GNA MOO.
WWW: http://130.212.41.61/moo.html
Telnet: baymoo.sfsu.edu 8888

ChibaMOO (the Sprawl) is the world's first public-access web server and WOO. ChibaMOO is loosely themed around the cyberscape of William Gibson's cyberpunk literature and now consists of several thousand rooms and nearly two thousand players and is growing at a rate of ten players a day. All objects on the Sprawl are accessible via MUD or WWW client, and anyone is welcome to publish their own non-commercial WWW page there. The Sprawl is sponsored by SenseMedia (http://sensemedia.net/).
WWW: http://sensemedia.net/sprawl/11
Telnet: chiba.picosof.com 7777

LambdaMOO describes itself as "a new kind of society, where thousands of people voluntarily come together from all over the world." The administrators warn that visitors may not always like what others have to say and advise newcomers "to be careful who you associate with and what you say." As one of the oldest, most diverse, and largest MOOs in existence, LambdaMOO is worth a visit.
Telnet: lambda.moo.mud.org 8888

Selected Educational Web Sites on MOOs and MUDs

Courses That Use Virtual Technology in Teaching, the PennMOO web site, offers a list of courses being taught with MOOs at the University of Pennsylvania and a short sampling of similar courses across the Internet. (See PennMOO above).
WWW: http://www.english.upenn.edu/PennMOO/related.html

Educational Technology: Educational VR (MUD) subpage is one of the most comprehensive educational sites for MOOs and MUDs. This site provides users from all experience levels information on events, what's new, general index and information pages, educational MUDs list, publications, bibliographies and document indexes and mailing lists, MOO/MUD guides, faqs and manuals, teaching in cyberspace, and teaching about cyberspace, clients and servers, 2/3d multi-user virtual worlds, and MUD research. This is a must-see web site.
WWW: http://tecfa.unige.ch/edu-comp/WWW-VL/eduVR-page.html#Teaching

The Evolving TecfaMOO Book might be thought of as an everything-you-might-want-to-know site about TECFA MOO, a virtual space for educational technology, education, research and life at TECFA, School of Psychology and Education, University of Geneva, Switzerland. The mission of this MOO is related to research in computer-mediated communication and to its use as communication tool for the diploma in educational technology and to student projects in the STAF-14 course. Of particular note in this evolving book is the section on MOOs for education, with sections on virtual classrooms and MOO learning environments.
WWW: http://tecfa.unige.ch/moo/book1/tm.html

MUD/MOO/MUSH Libraries

Amberyl's Almost-Complete List of MUSHes offers a personal list of MUSHes, together with addresses, commentary, links to home pages, and, if available, MudWHO queries. It is in alphabetical order and should be fairly complete.
WWW: http://www.clock.org/muds/muds.html

Foreign Language Resource Web: While Steve Thorne's noncomprehensive list of foreign-language resources has little to do with MOOs, the resource serves as a nice companion to the foreign-language MOOs and MUD listed above.
WWW: http://www.itp.berkeley.edu/~thorne/HumanResources.html

Freya's list o'moos is by far the most often updated and comprehensive list of MOOs available on the WWW. All of her MOOs are color-coded as either social/miscellaneous, role-playing game (rpg), language, or educational/research.
WWW: http://www.teleport.com/~autumn/moo.html

Gopher archives on Networked Virtual Realities provides a set of resources for those building educational environments using networked virtual-reality software such as MUSE, tinyMUSH, and MOO to pool knowledge and discuss issues. This site also archives the CBNVEE listserv. Some of the archived discussions include graphical MU*s, VR and Education Bibliography, Letter to sites banning MOOs, and forced interactivity. To subscribe to CBNVEE, send the following command to listserv@mcmuse.mc.maricopa.edu via electronic mail: subscribe cbnvee your fullname. For example: subscribe cbnvee John Doe.
gopher://mcmuse.mc.maricopa.edu:70/11/cbnvee

MOO/MU* Document Library is listed at the WWW Virtual Library of MOOs and MUDs site but is worth listing here again. It serves as one of the

definitive document libraries of MUD/MOO research available on the WWW.
WWW: http://lucien.berkeley.edu/moo.html

WWW virtual library of MOOs and MUDs is a component of The MUD/MUSH/MOO Catalog of Catalogs. Though this site lists only about a dozen URLs for other resources, they represent the biggest and most comprehensive MOO/MUD libraries available on the Internet. This site is a great starting point for general information and thorough research.
WWW: http://www.educ.kent.edu/mu/catofcat.html

Contributors

John F. Barber is Assistant Professor at Northwestern State University of Louisiana. He is interested in working at the intersection of teaching, learning, and computer technology to create collaborative virtual classrooms, and in the pedagogical issues raised through efforts to improve the teaching and learning of new literacies in these spaces.

Jorge R. Barrios is a student of informatics at the University of Oslo, Norway. He has been involved with MOOs since early 1994. He is the cofounder of Meridian, a MOO dedicated to virtual travel and cultural interaction, where he has designed a number of tools for research and education.

Mark Blanchard is a recent graduate in computer science at Buena Vista University in Storm Lake, Iowa. His current research is on compiler theory and parsing and the Java programming language. His experience with MOOs dates back several years, during which time he has been working with different ways to interface MOOs with the World Wide Web.

Amy Bruckman is Assistant Professor in the College of Computing at the Georgia Institute of Technology, where she does research on virtual communities and education. She received her Ph.D. from the Epistemology and Learning Group at the Media Lab at MIT in May 1997. Dr. Bruckman is the founder of MediaMoo, an online community for media researchers, and MOOSE Crossing, a MUD designed to be a constructionist learning environment for kids. She received her bachelor's degree in physics from Harvard University in 1987, and her masters from the MIT Media Lab's Interactive Cinema Group in 1991.

Juli Burk is Associate Professor of Theatre in the University of Hawaii Department of Theatre and Dance. She is the administrator of ATHEMOO, developed for the Association for Theatre in Higher Education (ATHE). She is a scholar and stage director whose specialties are contemporary and feminist theory.

Brian Clements is co-founder of Rancho Loco Press. His poems and essays have appeared in journals and magazines such as *Agni, American Literary Review, Another Chicago Magazine, Plum Review, Southwestern American Literature,* and the *Wallace Stevens Journal.* His poems also appear in two volumes, *Essays Against Ruin* (Texas Review Press, 1997) and *Flesh and Wood* (Mbira Press, 1992).

Eric Crump is the learning-technologies coordinator for the University of Missouri-Columbia Learning Center, where he is developing network-based writing and learning environments and other learning opportunities for faculty and students. On leave from UM-C, Crump is Internet Coordinator for the National Council of Teachers of English (NCTE). He is the founder and editor of *RhetNet,* a cyberjournal for rhetoric and writing, and is the managing editor of the *Wakonse Journal of College Teaching and Learning.*

Pavel Curtis received his bachelor's degree from Antioch College in 1981 and his masters and doctorate from Cornell University in 1983 and 1990, respectively. He was a member of the research community at the Xerox Palo Alto Research Center from 1983 through 1996, during which time he worked on aspects of the Smalltalk-80, Interlisp-D/Xerox Lisp, and Cedar programming environments and on other projects related to the design and implementation of programming languages. He was leader of the Scheme-Xerox project exploring large-scale software development in the Scheme programming language. He is the founder and chief administrator of LambdaMOO, one of the most popular recreational social virtual realities on the Internet. Since 1996, as a principal architect and cofounder of PlaceWare Inc., he has been working on bringing the benefits of network places to a broader range of users and applications.

D. Diane Davis is Assistant Professor of Rhetoric at the University of Iowa, where she teaches virtual rhetorics, rhetoric and writing theory, and contemporary critical theory. Her most recent work addresses the issue of community in a post-humanist age, specifically the ways in which the construction of electric subjects both re/produces what is called "the human" and challenges the boundaries of "humanism." Her work has appeared in *PRE/TEXT, Rhetoric Review, Rhetoric Society Quarterly, PRE/TEXT Electra(Lite), The Encyclopedia of Rhetoric, Studies in Psychoanalytic Theory,* and *Composition in Context.* Her book *Breaking Up [at] Totality: A Rhetoric of Laughter* is forthcoming from Southern Illinois University Press.

Jeffrey R. Galin is Assistant Professor of English, specializing in composition, at California State University, San Bernardino, where he teaches graduate and undergraduate composition, composition theory, and electronic pedagogy. He is currently coauthoring *The Dialogic Classroom: Teachers Integrating Computer Technology, Pedagogy, and Research.*

Dene Grigar is Assistant Professor of English at Texas Woman's University, where she teaches composition in a networked computer lab, as well as literature courses focusing on Greek epic and modernist poetry. She is currently collaborating on a series of dialogues, written online, that explores the effects of electronic technology upon academia, entitled *Four Dialogues: Ancient Philosophy and Modern Technology.*

Cynthia Haynes is Assistant Professor in the School of Arts and Humanities and Director of Rhetoric and Writing at the University of Texas at Dallas, where she teaches both graduate and undergraduate rhetoric, composition, and electronic pedagogy. Her publications have appeared in *PRE/TEXT, Composition Studies, Journal of Advanced Composition, Keywords in Composition, St. Martin's Guide to Tutoring Writing, Works and Days, Writing Center Journal, Kairos,* and *CWRL.* She is coeditor of *Pre/Text: Electra(Lite),* and with Jan Rune Holmevik, she is co-founder of Lingua MOO. Her interests in rhetorical delivery and electronic scholarship spawned the C-FEST series of online real-time meetings at Lingua MOO, where these issues are debated year-round. She is currently at work on *Technologies of Ethos: Rhetoric and the New Delivery of the Humanities.*

Jan Rune Holmevik is a visiting assistant professor and doctoral candidate in the Department of Humanistic Informatics at the University of Bergen, Norway. He holds a Cand. Philol. degree in the history of technology from the University of Trondheim, Norway (1994), and his publications on history of computing and science policy have appeared in journals such as *Annals of the History of Computing* and *Forskningspolitikk.* He is coauthor of *MOOniversity: Students and the On-line Classroom* (forthcoming Allyn and Bacon, 1998) with Cynthia Haynes, and author of *The Digital Factor* (forthcoming from Ad Notam Gyldendal). Holmevik has been involved with MUDs since 1989 and is cofounder of Lingua MOO.

Michael Joyce is Professor of English at Vassar College. He is the author of *Afternoon, a story,* perhaps the most celebrated hypertext fiction written to date, and coauthor of the hypertext authoring software Storyspace. He has lectured and published extensively on issues relating to hypertext and writing, and his most recent collection of essays, *Of Two Minds: Hypertext Pedagogy and Poetics,* was published by the University of Michigan Press in 1994.

Beth Kolko is Assistant Professor of English at the University of Texas at Arlington, where she teaches courses on electronic discourse, technology and culture, and composition. Her recent publications have focused on rhetoric, narrative, and virtual communities. Her research more generally investigates writers outside of educational environments, and she is currently working on a longer project that examines the social nature of writing and narrative collaboration in virtual communities.

Ken Schweller is Professor of Computer Science and Psychology at Buena Vista University in Storm Lake, Iowa. He has published in the areas of speech act theory and computer science, and his research interests include artificial intelligence, virtual communities, and Java network programming. He is the founder and head administrator of CollegeTown MOO.

Sherry Turkle is Professor of the Sociology of Science at the Massachusetts Institute of Technology. She has written numerous articles on psychoanalysis and culture and on the "subjective side" of human relationships with technology, especially computers. She is the author of *Psychoanalytic Politics: Jacques Lacan and Freud's French Revolution* (1978) and *The Second Self: Computers and the Human Spirit* (1984). Her most recent research is on the psychology of computer-mediated communication, including role-playing on MUDs, which is reported in her latest book, *Life on the Screen: Identity in the Age of the Internet* (1995).

Victor J. Vitanza is Professor of English at the University of Texas at Arlington. His work includes rhetorical theories, histories of rhetoric, composition theories, electronic pedagogy, and Third Sophistic subjectivities. He is the founder and editor of *PRE/TEXT: A Journal of Rhetorical Theory* and coeditor of *PRE/TEXT: Electra(Lite).* He is also the moderator of the PRETEXT List, on Rhetorical Theory, at the Spoon Collective, and Editor of *CyberReader* (1996). His newest book is *Negation, Subjectivity, and the History of Rhetoric* (SUNY Press).

Shawn P. Wilbur is a doctoral candidate in American Culture Studies at Bowling Green State University, where he studies the implications of "virtual community" as a metaphor for online collectivity. He is a member of the Spoon Collective, which maintains a group of more than twenty online philosophy discussion lists, editor of the electronic magazine *Voices from the Net,* and an administrator of PMC2 (previously Postmodern Culture MOO).

Deanna Wilkes-Gibbs is a research scientist at Bellcore, where she studies the use of network communications technologies for facilitating interpersonal and collaborative interaction at a distance. She holds a doctorate in cognitive psychology from Stanford University and has written and published widely on the psychology of language use, especially with regard to the pragmatics of conversational interaction. She is cofounder of the education-related MOO Meridian.

UNIVERSITY OF MA
3 2304 00064815 2